Users Guide to Physical Modelling and Experimentation

T0362046

IAHR Design Manual

Series editor

Peter A Davies
Department of Civil Engineering,
The University of Dundee,
Dundee,
United Kingdom

The International Association for Hydro-Environment Engineering and Research (IAHR), founded in 1935, is a worldwide independent organisation of engineers and water specialists working in fields related to hydraulics and its practical application. Activities range from river and maritime hydraulics to water resources development and eco-hydraulics, through to ice engineering, hydroinformatics and continuing education and training. IAHR stimulates and promotes both research and its application, and, by doing so, strives to contribute to sustainable development, the optimisation of world water resources management and industrial flow processes. IAHR accomplishes its goals by a wide variety of member activities including; the establishment of working groups, congresses, specialty conferences, workshops, short courses; the commissioning and publication of journals, monographs and edited conference proceedings; involvement in international programmes such as UNESCO, WMO, IDNDR, GWP, ICSU, The World Water Forum; and by co-operation with other water-related (inter)national organisations.
www.iahr.org

Supported by
CEDEX

Users Guide to Physical Modelling and Experimentation

Experience of the HYDRALAB Network

Editors

L.E. Frostick, S.J. McLelland & T.G. Mercer

Department of Geography, University of Hull, Hull, UK

Lead authors

J. Kirkegaard, G. Wolters, J. Sutherland, R. Soulsby, L. Frostick, S. McLelland, T. Mercer & H. Gerritsen

CRC Press
Taylor & Francis Group
Boca Raton London New York Leiden

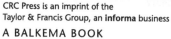

CRC Press is an imprint of the
Taylor & Francis Group, an **informa** business

A BALKEMA BOOK

Acknowledgement:
The work described in this publication was supported by the European Community's Sixth Framework Programme through the grant to the budget of the Integrated Infrastructure Initiative HYDRALAB III, Contract no. 022441 (RII3).

Please refer to this publication as follows:
Frostick, L.E.; S. J. McLelland & T.G. Mercer (eds), 2011. Users Guide to Physical Modelling and Experimentation: Experience of the HYDRALAB Network, CRC Press/Balkema, Leiden, The Netherlands.

Typeset by Vikatan Publishing Solutions (P) Ltd, Chennai, India
Printed and bound in Great Britain by Antony Rowe (A CPI-group Company), Chippenham, Wiltshire

Published by: CRC Press/Balkema
P.O. Box 447, 2300 AK Leiden, The Netherlands
e-mail: Pub.NL@taylorandfrancis.com
www.crcpress.com – www.taylorandfrancis.co.uk – www.balkema.nl

CRC Press/Balkema is an imprint of the Taylor & Francis Group, an informa business

Library of Congress Cataloging-in-Publication Data
Users Guide to Physical Modelling and Experimentation: Experience of the HYDRALAB Network / editors, L.E. Frostick, S.J. McLelland & T.G. Mercer. —1st ed.
p. cm.
Includes bibliographical references and index.
ISBN 978-0-415-60912-8 (pbk. : alk. paper)—ISBN 978-1-4398-7051-8 (e-book)
1. Hydraulic models. 2. Hydraulic engineering—Mathematical models.
3. Hydrodynamics—Mathematical models. I. Frostick, L.E. II. McLelland, S.J. (Stuart J.) III. Mercer, T.G. (Theresa G.) IV. Title: Users guide to hydraulic modelling and experimentation.

TC164.H915 2011
627.01'1--dc22

2010054312

ISBN: 978-0-415-60912-8 (Pbk)
ISBN: 978-1-4398-7051-8 (eBook)

About the IAHR Book Series

An important function of any large international organisation representing the research, educational and practical components of its wide and varied membership is to disseminate the best elements of its discipline through learned works, specialised research publications and timely reviews. IAHR is particularly well-served in this regard by its flagship journals and by the extensive and wide body of substantive historical and reflective books that have been published through its auspices over the years. The IAHR Book Series is an initiative of IAHR, in partnership with CRCPress/Balkema - Taylor & Francis Group, aimed at presenting the state-of-the-art in themes relating to all areas of hydro-environment engineering and research.

The Book Series will assist researchers and professionals working in research and practice by bridging the knowledge gap and by improving knowledge transfer among groups involved in research, education and development. This Book Series includes Design Manuals and Monographs. The Design Manuals contain practical works, theory applied to practice based on multi-authors' work; the Monographs cover reference works, theoretical and state of the art works.

The first and one of the most successful IAHR publications was the influential book *"Turbulence Models and their Application in Hydraulics"* by W. Rodi, first published in 1984 by Balkema. I. Nezu's book *"Turbulence in Open Channel Flows"*, also published by Balkema (in 1993), had an important impact on the field and, during the period 2000–2010, further authoritative texts (published directly by IAHR) included *Fluvial Hydraulics* by S. Yalin and A. Da Silva and *Hydraulicians in Europe* by W Hager. All of these publications continue to strengthen the reach of IAHR and to serve as important intellectual reference points for the Association.

Since 2011, the Book Series is once again a partnership between CRCPress/Balkema - Taylor & Francis Group and the Technical Committees of IAHR and I look forward to helping bring to the global hydro-environment engineering and research an exciting set of reference books that showcase the expertise within the IAHR Community.

Peter A Davies
University of Dundee, UK
(Series Editor)

Contents

Foreword

In hydrodynamics research the roles of physical modelling and laboratory experimentation (the respective terms adopted mostly for the same activity in the hydraulics and fluid dynamics communities) have undergone important changes in recent years. For a considerable period of time during the last 20–30 years, the activity was not only slightly unfashionable but also lacking somewhat in impact so far as important advances in the subject were concerned. Important, site-specific physical modelling investigations related to infrastructure projects (ports, harbours, breakwaters) continued to be undertaken but controlled parametric modelling studies fell out of favour so far as hydrodynamics research was concerned. Numerical modelling approaches seemed to be much more attractive to researchers and practitioners. So what has changed to make their value more crucially important? Firstly, the development of sophisticated equipment and instrumentation systems have provided opportunities for measuring three-dimensional, unsteady turbulent motion and concentration fields to levels of precision that were previously inaccessible. The associated development of experimental techniques (particularly those that are non-intrusive) together with advances in automated data acquisition and data analysis systems, rapid processing and increased data storage capabilities has provided measurements having temporal and spatial resolution that make them directly useful for validating predictions from advanced numerical models. The individual chapters in this Handbook illustrate the enormous range of experimental tools available for specific applications. The transition to the present arrangements seen in most leading laboratories of computer-controlled, automated, reliable systems for point and synoptic flow and concentration field measurements has been remarkable.

The other important development has been the demonstration of the quantitative deficiencies and limitations of predictive numerical models when applied in isolation to complex flows. Many of these deficiencies were only apparent when CFD models, in particular, were evaluated in test case comparisons. Such findings exposed the risks of reliance on numerical models in isolation.

Against this background, far-sighted individuals and institutions made significant investments in laboratory facilities, buildings, equipment and technical expertise, in order to respond to the emerging research need for physical model data. In Europe, the *HYDRALAB* Programmes played a key role in maximising the potential of this investment and harmonising the capabilities of each large facility that had been developed. This process has been enormously successful and productive, with benefits extending far beyond the European context. The founders of *HYDRALAB* recognised at the outset that, although empirically-based research has its place, the emphasis should be to encourage the development of a theoretical framework for each hydrodynamics project supported. The *HYDRALAB* Programmes (one of which – *HYDRALAB III* is the primary focus of this Handbook) led the way in demonstrating the need for synergy between the various research tools (physical models, numerical and analytical models and field experiments) to tackle the flow problems at the forefront of fluid mechanics and hydraulics research. Such problems are now often so complex that only an integrated approach is feasible if the flow is to be described sufficiently well for practitioners to use the results with confidence. In areas where an empirical approach was the norm, *HYDRALAB* has commendably introduced measures to force the inclusion of fundamental understanding into all its sponsored investigations, to demonstrate that research is not simply a data collection exercise. The laboratory data inform the theoretical model development and the theoretical model generates questions requiring laboratory intervention and investigation. The promotion of the *CoMIBBS* joint research project within *HYDRALAB III* is a clear recognition of this approach.

A Users Guide to Physical Modelling and Experimentation is an ambitious title but it provides (i) a systematic, comprehensive summary of the lessons learnt in *HYDRALAB III*, (ii) recommendations for the best practice to be adopted when carrying out state-of-the-art, modern laboratory investigations and (iii) an inventory and review of the recent advances in instrumentation and equipment that drive present and new developments in the subject. The *Guide* concentrates on four core areas – waves, breakwaters, sediments and the relatively-new (but rapidly-developing) cross-disciplinary area of hydrodynamics/ ecology. The experiences gained in the *CoMIBBS* component of *HYDRALAB III* provide the material for the chapter on composite modelling guidance, principles and practice. Much of the detailed consideration of scaling and the degree of relevance of laboratory/physical modelling approaches is given for specific contexts in the individual chapters. The *Guide* includes outputs from the many transnational access projects that have been supported within *HYDRALAB III*, as well as the focussed joint research activities *SANDS* and *CoMIBBS*. Its primary purpose is to serve as a shared resource containing the outstanding advances achieved within *HYDRALAB III* but, even more than this, it incorporates the human and institutional collaborations that led to and sustained the research advances, the human relationships that were strengthened and initiated through joint participation in the Programme and the training opportunities that participation provided to the many young researchers engaged on the projects.

The publication of this User Guide is extremely timely. In Summer 2009 at the 33rd Congress of the International Association for Hydro-Environment Engineering & Research (IAHR), *HYDRALAB III* led a Special Seminar "Experimental research around the world" not only to showcase the approach adopted by the Programme but also to gauge the extent of interest outside Europe in this field. It was clear that the developments exploited within *HYDRALAB III* had much in common with emerging research trends and practice elsewhere.

This User Guide disseminates an enormous amount of valuable advice and information gathered within *HYDRALAB III* and it constitutes a resource that will not only provide a record of the accomplishments of the Programme but also serve as an authoritative text that will have applicability and utility well beyond the European context covered by *HYDRALAB III*. I have great pleasure in recommending it.

Peter A. Davies
Chair, HYDRALAB III *International Advisory Board*

Acronyms and Abbreviations

2DV	Two dimensions in vertical
2DH	Two dimensions in horizontal
3D RANS	Three dimensional Reynolds Averaged Navier-Stokes
ABS	Acoustic Backscatter System
ADV	Acoustic doppler Velocimeters
ADVP	Coherent Acoustic Doppler Profile
AOFT	Aberdeen Oscillating Flow Tunnel
AUKE	Wave generation and postprocessing software used at Deltares – Delft Hydraulics
BC	Boundary Conditions
BDM	Bayesian Directional Method
BSS	Brier Skills Scores
CCM	Conductivity Concentration Meter
CEH	Centre for Ecology and Hydrology
CFD	Computational fluid dynamics
CIEM	Flume at UPC
CM	Composite Modelling
CoMIBBS	Composite Modelling for the Interactions Between Beaches and Structures
CSD	Computational Structure Dynamics
DANS	Double-averaged continuity and Navier-Stokes equations
DEM	Digital Elevation Model
DH	Deltares – Delft Hydraulics
DHI	DHI
DTU	Technical University of Denmark
EU	European Union
GWK	Großer Wellenkanal
HWA	Hot Wire Anemometry
JONSWAP	Joint North Sea Wave Project
LCS	Low crested structures
LAI	Leaf Area Index
LDA	Laser Doppler Anemometry
LDV	Laser Doppler Velocimeter
LIF	Laser Induced Fluorescence
LOWT	Large Oscillating Water Tunnel

LNEC	Laboratório Nacional de Engenharia Civil
LTT	Low Tide Terrace
MEM	Maximum Entropy Method
MLM	Maximum Likelihood Method
MMLM	Modified Maximum Likelihood Method
MLP	Multi Layer Perceptron
NERC	Natural Environment Research Council
NM	Numerical Model
NN	Neural Network
NTNU	Norwegian University of Science and Technology
OBS	Optical Backscatter Sensor
OWT	Oscillating Water Tunnels
PISCES	Coastal area model
PIV	Particle Image Velocimetry
PM	Physical Model
POLIBA	Politecnico di Bari
PTV	Particle Tracking Velocimetry
QA	Quality Assurance
Q3D	Quasi three dimensional
RANS	Reynolds-averaged continuity and Navier-Stokes equations
RMS	Root Mean Square
RP	Ripple Profiler
RVR	Root Volume Ratio
SANDFLOW	Sand transport model
SST	Sheer-Stress Transport
SWAN	Third generation coastal area wave model
TELEMAC	Finite element flow model
TES	Total Environmental Simulator
TKE	Turbulent Kinetic Energy
TMA	Texel-Marsen-Arsloe wave spectrum
UHCM	Ultra High Concentration Meter
UC	University of Catania
UHANN/FZK	Universitaet Hannover, Forxchungzentrum Küste
UPC	University Polytechnic of Catalunya
UT	University of Twente
UU	University of Utrecht
VLPM	Very Large Physical Model

List of symbols

Symbol	Units	Meaning
a_1	m	Length of the Low Tide Terrace (LTT)
a_2	m	Height of the Low Tide Terrace (LTT)
ak^2		Wave amplitude multiplied by wave number squared
A	m	Amplitude of near-bed wave orbital excursion
A_e	m²	Cross-sectional erosion area on rock profile
A_p		Position accuracy
b	m	Flume/Channel width
B	m	Width of test section
B_E		Transport error term
C	m$^{1/2}$/s	Chézy coefficient
C_d		Drag coefficient corresponding to a single plant
C_s	g/l	Sediment concentration
C_{SE}		Sediment budget error
Ca		Cauchy number
d	m	Particle size
$d*$		Non-dimensional particle diameter
D	m	Diameter of monopile, pipeline etc
$D*$		Non-dimensional grain diameter
D_{50}	m	Median nominal diameter of sediment or stone for which 50% of the total sample mass is lighter
D_{min}	m	Minimum diameter
D_g	m	Grain/Rock size
D_n	m	Nominal diameter of stone
D_{NO}		Normalisation term
D_S	m	Diameter of Structure
D_{Trunk}	m	Average trunk diameter
D_v	m²/s	Vertical turbulent diffusion coefficient
D_w		Dean parameter
e		Vogel exponent

E	N/m²	Modulus of elasticity
f	Hz	Wave frequency
f_m		Ripples in model
f_p		Flat bed in prototype
f_R		Reduced frequency
F	N	Force
F^*		Shields parameter
F_e	N	Elastic compression force
F_g	N	Gravity force
F_i	N	Inertia force
F_p	N	Pressure force
F_s	N	Surface tension force
F_i	N	Fluid friction force
Fd	N	Drag force
Fr		Froude number
g	m/s²	Gravitational acceleration
h	m	Water depth
h_c	m	Closure depth
h_{plant}	m	Tree height
H	m	Wave height
H_b	m	Wave breaking height
$H_{1/3}$	m	Wave height of the highest 1/3 of waves
$H_{1/10}$	m	Wave height of the highest 1/10 of waves
H_{m0}	m	Spectral significant wave height
H_s	m	Significant wave height
i, j		Standard tensor notation
k		Turbulent kinetic energy
$\langle k \rangle$	m²/s²	Spatially-averaged turbulent kinetic energy
k_l	m	Relevant length scale of organism
k_s	m	Nikuradse bed roughness
k_w	m⁻¹	Wave-number
KC		Keulegan-Carpenter number
l	m	Length
L	m	Characteristic length or local wave length
L_0	m	Deep-water wave length
Lb		Wave breaking position
L_p		Peak wave length
L_s		Larger surrogate
L_T		Wave load targets
L_{Trunk}	m	Average Trunk Length
m	kg	Mass

m_n	m^2/s^n	Spectral moments
Mb	kg	Body mass
M_{50}	kg	Mass for which 50% of the total sample mass is of lighter stones
n		Scale factor
N		Number of waves
N_d	%	Damage percentage
N_h		Geometric model scale factor in vertical direction, and for depths. $N_h = N_l$ in undistorted models.
N_l		Geometric model scale factor (in horizontal direction, for distorted models)
N_{od}		Number of displaced units within a strip with width of $1D_n$
N_S		Stability number; $N_S = H_S/\Delta D_{n50}$
N_X		Model scale factor – ratio of prototype value of a physical parameter X to model value of X, e.g. N_t, N_{ws}, N_{qb} are scale factors for t, w_s, q_b
$N_{displaced}$		Total number of stones that are displaced more than one diameter
N_{total}		Total number of stones within a section/layer
N_{vol}		Volumetric scale for the sediment
Nov		Outward unit vector normal to the bed surface
p'	Pa	Pressure fluctuation
P	Pa	Pressure
P_k	m^2/s^2	Production of turbulent kinetic energy
q	m^3/s	Mean overtopping discharge
q_b		Sediment/bedload transport rate
q_s	g	Suspended load
q_{tot}	g	Total load transport
q_t	m^2/s	Total sediment transport rate (volumetric)
Q		Transport rates
Q_b	m^3/s	Longshore bedload sediment transport rate integrated across surf zone (volumetric)
Q_f	$kg/m^2/s$	Net flux of sediments into suspension
Q_s	m^3/s	Suspended sediment transport rate integrated across a river or flume cross-section (volumetric)
R		Reflection coefficient
R_B		Branching ratio
R_D		Diameter ratio
Re		Reynolds number
$Re*$		Grain Reynolds number
Re_D		Reynolds number for flow conditions in the armour layer
R_L		Length ratio
s		Relative density of sediment

S		Equilibrium scour depth
S_c		Scour depth
S_d		Damage level parameter
S_f		Sheltering factor
$S_i(f)$	m²/Hz	Energy density of incident waves
$S_r(f)$	m²/Hz	Energy density of reflected waves
Sl		Slope of bed
Swb		Extent of the water-bed interface bounded by the averaging domain
t	s	Time
t_b		Bottom shear stress
T		Wave period
$T_{1/3}$	s	Wave period of the highest 1/3 of waves
T_{01}	s	Spectral mean period ($T_{01} = m_0/m_1$)
T_{02}	s	Spectral mean period ($T_{02} = (m_0/m_2)1/2$)
T_m	s	Mean wave period
T_M	s	Morphological timescale
$T_{m-1,0}$	s	Spectral wave period ($T_{m-1,0} = m_{-1}/m_0$)
T_n		Wave number
T_p	s	Peak wave period
T_s	s	Length of time series
T_w	s	Time for a wave to travel to the reflector and back
T_z	s	Average zero crossing wave period
u	m/s	Horizontal particle velocity
u'		Root mean square value of the fluctuating component of streamwise velocity
$u*$	m/s	Friction velocity
u'_i	m/s	Fluctuating part of the velocity
$\tilde{u}_i$	m/s	Spatially-varying part of the time-averaged velocity
$\langle \overline{u}_i \rangle$	m/s	Time-space averaged velocity
$\langle u'_i \rangle$		Space averaged temporal fluctuation
u''_i		Spatial deviation in time fluctuation
U		Ursell parameter
U_c		Current velocity
U_{cr}	m/s	Threshold velocity for sediment motion
U_{cw}	m/s	Relative velocity
U_f	m/s	Flow velocity (general)
U_i	m/s	Mean velocity flow (Navier-Stokes equation)
U_w	m/s	Orbital velocity
v	m/s	Velocity
$v*$	m/s	Shear velocity

v_o	m/s	Orbital velocity
v_{max}	m/s	Maximum wave velocity
v_{rms}	m/s	Root mean square velocity
V	m/s	Free-stream velocity
V_0	l	Total volume of the averaging domain
V_f	l	Fluid-only volume
w_s	m/s	Settling (fall) velocity of sediment grains (positive downwards) in quiescent water
We		Weber number
x		Constant for only wave situation
x_b	m	Distance inshore from position of wave breaking
x, y, z	m	Right-handed Cartesian coordinate system: x, horizontal axis, positive down flume or towards beach; z, vertical axis relative to bed level or a horizontal datum level; y, mutually orthogonal to x and z
Z	–	Suspension parameter
α	°	Angle of foreshore slope
β	°	Angle of wave attack with respect to the structure
Δ		Relative density: $(\rho a - \rho w)/\rho w$,
δ	m	Boundary layer thickness
δ^*	m	Displacement thickness
δ_{ij}		Kronecker delta function
$\langle \varepsilon \rangle$	m²/s³	Spatially averaged turbulent dissipation rate
η	m	Water surface elevation
η_k		Efficiency of the production of turbulent energy
η_ε		Efficiency of dissipation of turbulent energy
θ		Shields parameter
ϑ_{cr}		Threshold Shields parameter
μ		Ripple factor
μ_v	Pa s	Viscosity
ρ	kg/m³	Density
ρ_a	kg/m³	Mass density of rock armour
ρ_f	kg/m³	Mass density of fluid
ρ_s	kg/m³	Mass density of sediment
ρ_w	kg/m³	Mass density of water
σ	N/m	Surface tension
σ_a		Form parameter of the JONSWAP spectrum
σ_b		Form parameter of the JONSWAP spectrum
σ_{ij}	N/m²	Stress fluctuations
τ	N/m²	Bed shear stress

τ_b	N/m²	Bottom shear stress
τ_{cr}	N/m²	Threshold bed shear stress for sediment motion
τ_{cs}	N/m²	Sheer stress at which suspended load is initiated
τ_{ij}	N/m²	Normal pressure stresses
υ_k	m²/s	Kinematic viscosity of water
$\upsilon_{t\phi}$	m²/s	Turbulent eddy viscosity
γ		Peak enhancement factor of the JONSWAP spectrum
ϕ		Porosity
ϕ_s		Spatial porosity
ϕ_t		Time porosity

Contributors

HYDRALAB
Partner/Invited Participants
   Contributor   **E-Mail**

Centre for Ecology and Hydrology (CEH)
   Dr. Pamela Naden   psn@ceh.ac.uk
   Dr. Ponnambalam Rameshwaran   ponr@ceh.ac.uk

CNRS
   Dr. Joel Sommeria   sommeria@coriolis-legi.org

Deltares | Delft Hydraulics (DH)
   Mark Klein Breteler   Mark.KleinBreteler@deltares.nl
   Dr. Henk Van Den Boogaard   henk.vandenboogaard@deltares.nl
   Dr. Sofia Caires   sofia.caires@deltares.nl
   Neelke Doorn   n.doorn@tudelft.nl
   Dr. Herman Gerritsen   Herman.Gerritsen@deltares.nl
   Ir. Ellis Penning   ellis.pennning@deltares.nl
   Dr. Marcel van Gent   marcel.vangent@deltares.nl
   Prof. Leo van Rijn   leo.vanrijn@deltares.nl
   Dr. Guido Wolters   guido.wolters@deltares.nl

DHI
   Rolf Deigaard   rd@dhigroup.com
   Nicholas Grunnet   ngr@dhigroup.com
   Jens Kirkegaard   jkj@dhigroup.com
   Doris Mühlestein   dm@dhigroup.com
   Hemming Schäffer   Hemming@SchafferWaves.dk

Technical University of Denmark (DTU)
   Martin Dixen   mdi@dhigroup.com
   Jørgen Fredsøe   jf@mek.dtu.dk
   Palle Martin Jensen   pmje@mek.dtu.dk
   Prof. B. Mutlu Sumer   bms@mek.dtu.dk

HR Wallingford Ltd.
   William Allsop   w.allsop@hrwallingford.co.uk
   Roger Bettes   r.bettess@hrwallingford.co.uk

Dr. Charlotte Obhrai   cob@hrwallingford.co.uk
Prof. Richard Soulsby   r.soulsby@hrwallingford.co.uk
Dr. James Sutherland   j.sutherland@hrwallingford.co.uk

Laboratório Nacional de Engenharia Civil (LNEC)
Rui Capitão   rcapitao@lnec.pt
Juana Fortes   jfortes@lnec.pt
Rute Lemos   rlemos@lnec.pt
Dr. Paula Freire   pfreire@lnec.pt
Dr. Maria da Graça Neves   gneves@lnec.pt
Dr. Filipa Oliveira   foliveira@lnec.pt
Artur Palha   aclerigo@lnec.pt
Eng. Liliana Pinheiro   lpinheiro@lnec.pt
Maria Teresa Reis   treis@lnec.pt
Manuel Marcos Rita   mrita@lnec.pt
Dr. Francisco Sancho   fsancho@lnec.pt
Dr. João Santos   jasantos@lnec.pt
Isaac Sousa   isousa@lnec.pt

Loughborough University
Dr. Stephen Rice   S.Rice@lboro.ac.uk

NERC (Proudman Oceanographic Laboratory)
Prof. Peter Thorne   pdt@pol.ac.uk

Norwegian University of Science and Technology (NTNU)
Dr. Morten Alver   alver@itk.ntnu.no
Dr. Ingrid Ellingsen   Ingrid.Ellingsen@sintef.no
Alexandra Neyts   alexandra.neyts@bio.ntnu.no
Dr. Lasse Olsen   lasse.olsen@bio.ntnu.no

Politecnico di Bari (Poliba)
Prof. Leonardo Damiani   l.damiani@poliba.it
Dr. Francesca de Serio   f.deserio@poliba.it

SINTEF Marine
Carl Trygve Stansberg   Carl.Stansberg@marintek.sintef.no

SOGREAH Consultants
Dr. Luc Hamm   luc.hamm@sogreah.fr
Laurent Bonthoux   laurent.bonthoux@sogreah.fr

Total Environmental Simulator, University of Hull (TES)
Prof. Lynne Frostick   L.E.Frostick@hull.ac.uk
Dr. Stuart McLelland   S.J.McLelland@hull.ac.uk
Theresa Mercer   T.Mercer@hull.ac.uk

University of Catania (UC)
Prof. Enrico Foti   efoti@dica.unict.it
Dr. Rosaria E Musumeci   rmusume@dica.unict.it

Universitaet Hannover, Forschungzentrum Küste (UHANN/FZK)

Dipl.-Ing. Joachim Gruene   gruene@fzk.uni-hannover.de
Prof. Hocine Oumeraci   H.Oumeraci@tu-bs.de
Dr. Stefan Schimmels   schimmels@fzk.uni-hannover.de
Ulrike Schmidtke   prepernau@fzk.uni-hannover.de
Dipl.-Ing. Reinold Schmidt-Koppenhagen   sk@fzk.uni-hannover.de

Universitat Politecnica de Catalunya (UPC)

Dr. Iván Cáceres   i.caceres@upc.edu
Dr. F. Xavier Gironella   xavi.gironella@upc.edu
Prof. Agustin Sanchez-Arcilla   agustin.arcilla@upc.edu
Dr. Joan Pau Sierra   joan.pau.sierra@upc.edu
Tiago Oliveira   tiago.oliveira@upc.edu

University of Twente (UT)

Dr. Jan Ribberink   j.s.ribberink@utwente.nl

University of Utrecht (UU)

Dr. Maarten Kleinhans   m.kleinhans@geo.uu.nl

VITUKI

László Rákóczi   lrakoczi@t-online.hu
György Szepessy   szepessy@vituki.hu

Chapter 1

Introduction

Lynne Frostick, Stuart McLelland & Theresa Mercer

J. Kirkegaard, G. Wolters, J. Sutherland, R. Soulsby,
L. Frostick, S. McLelland, T. Mercer & H. Gerritsen

1.1 INTRODUCTION

Physical modelling is an established technique for hydraulic research. It bridges the gap between what can be simulated accurately using numerical models and the real world. It also facilitates the calibration of numerical models and increases confidence in future predictions. These are vital given the present need to understand and adapt to the impacts of climate change. Worldwide it has been estimated that more than 2.75 billion people will live within 95 km of the coasts by the year 2025 (Science Daily, July 18 2006, *Center for Climate Systems Research*). Added to this is the impact of history that dictates that the majority of cities and towns are situated on rivers which have long been prime sources of water, waste disposal, food and transport. This growing section of the population are susceptible to the effects of changes in both rainfall patterns and sea level which control both the frequency and scale of river and coastal flooding and the rates of coastal erosion. It is therefore imperative that we find ways of predicting the consequences of future changes in river and coastal systems. This can only be achieved through modelling – and most predictions are now based on numerical models. However, such models are only as good as assumptions and data on which they are based. Physical models are one of the most cost-effective ways of providing these data. Physical hydraulic models which are well founded and controlled laboratory experiments simulating hydraulic systems are therefore important to both our present and future research needs. This book is designed to act as a guide to best practice in this type of laboratory experimentation. It encompasses work carried out during the Integrated Infrastructure Initiative project HYDRALAB III of the European Community's Sixth Framework Programme.

1.2 RATIONALE FOR THE BOOK

Most established laboratories that perform physical modelling have their own particular methods for conducting model studies. Some of these techniques have evolved over time as a result of experience, while others have been acquired through contact with laboratories doing similar types of modelling. This has resulted in physical modelling procedures for similar experiments in different laboratories varying considerably e.g. scaling methods, sediment recirculation, flow measurement, analysis and verification procedures etc. However, this makes the comparison of model results and the data transfer between varying laboratories very difficult. It also confuses new users of physical modelling facilities and does nothing to ensure that the experimental design finally adopted by these users is fit for purpose. The rationale for this book is therefore to disseminate knowledge, methodologies, instrumentation and practices among the physical modelling community, to become the basis for a unified approach to physical hydraulic modelling and to act as a reference for researchers new to the field. It will help simplify data transfer and the interpretation of modelling results gathered using different facilities and contrasting modelling approaches. It incorporates chapters on well developed fields of experimentation such as wave, breakwater and sediment research as well as one concerned with the newer and less well developed field of eco-hydraulic experimentation. The latter is an emerging field for hydraulic facilities and it introduces experimental complexities that are at the forefront of physical hydraulic modelling research.

1.3   ADVANTAGES AND DISADVANTAGES OF PHYSICAL MODELLING

There are many advantages of using physical models. They allow insight into phenomena that are not yet described or are so complicated that they are inaccessible to numerical modelling. They can also integrate the forcing physical processes without the simplifying assumptions that have to be made for analytical or numerical models. Physical models can be used to obtain measurements to validate and calibrate numerical models and for predicting the consequences of extreme conditions that cannot be measured in the field.

A further advantage of using physical models is the high degree of experimental control that allows simulation of varied and rare environmental conditions at the convenience of the researcher. Such models can also provide an immediate visual feedback of the physical processes which in turn can help to focus and modify the design of the study.

However there are some disadvantages associated with physical model testing that are related to experimenting in facilities of limited spatial extent which require that phenomena are scaled down from the natural system and which can, by their design, produce laboratory effects not present in nature. Scale effects occur in models that are smaller than the prototype where it is not possible to simulate all relevant variables in correct relationship to each other. Laboratory effects induced by model boundaries and unrealistic forcing conditions can influence the process being simulated. It should also be kept in mind that an incorrectly designed model always provides inaccurate predictions, independent of the sophistication of the instrumentation and measuring methods (Yalin, 1989). Despite these shortcomings, scale models in a hydraulic facility are often the best tool for engineers to discover and verify engineering solutions and to improve upon and verify mathematical and numerical approaches.

Laboratory models will always represent the full-scale problem with some simplification. Thus, the modeller is forced to compromise on the details of the model – simplify but do not over-simplify. The advantages and disadvantages of

Table 1.1 Advantages and disadvantages of physical modelling.

Advantages	Disadvantages
Allow insight into phenomena not yet described or understood	Scale effects occur in down-scaled models
Integrate the governing physical processes without simplifying assumptions that have to be made for analytical or numerical models	It is not possible to simulate all the relevant variables in correct relationship to each other
Can be used to obtain measurements to verify or disprove theoretical results	Laboratory effects induced by model boundaries and unrealistic forcing conditions can influence the process being simulated
Can be used to obtain measurements for phenomena so complicated that so far they have not been accessible for theoretical approaches	Facilities vary widely and are expensive to install and maintain. The correct facility for the application may not be available (see below on selection of facilities)
Can be used to obtain measurements for extreme conditions not measured in the field	
Give a high degree of experimental control that allows simulation of varied or sometimes rare environmental conditions at the convenience of the researcher	
Give a visual feedback from the model: The physical model provides an immediate qualitative impression of the physical processes which in turn can help to focus the study and reduce the planned testing	

physical model testing have already been laid out in a previous HYDRALAB report (HYDRALAB 2004) and these are summarised in Table 1.1 above.

The cost of physical modelling is often more than that of numerical modelling, but less than that of major field experiments. However the extent of the savings over field data collection depends on the exact nature of the problem being studied.

1.4   AUDIENCE

The book is aimed at a wide range of groups who might be involved with experimental physical modelling. These include:

- New recruits at hydraulic laboratories, as an introduction to physical modelling;
- More established staff at hydraulic laboratories, as a reference work and an insight into methods used at other laboratories;
- Clients of hydraulic laboratories, as a supporting document codifying the methodologies currently in use and their strengths and weaknesses;
- Researchers at universities, for many of the above reasons.

However, the book is *not* intended to be a student-level textbook.

1.5   CHOICE OF FACILITY

The choice of facility for a particular programme of research is the vital first step since a mistake at this stage may render the results useless. Facility selection will depend on the scale and scope of the phenomenon to be modelled. For example the prime interest may be in detail of processes and the size of the facility may not be critical, but the range of measuring equipment available may be. On the other hand, if the intention is to model flow-boundary interactions over a large area (such as in the modelling of wave response to breakwaters and installations) a large facility is essential so that model scales of 1:10 or 1:100 can be used. Combining detailed process and scale modelling in different facilities can prove very valuable, nesting the results of process studies of critical zones within those from a less detailed, broader scale model.

The range of model facility types can be divided into 'standard' facilities which are widely distributed (though of varying size), and specially designed facilities which are only found in a few laboratories in Europe. Standard facilities include current flumes, wave flumes, wave basins (with or without currents), river physical models and tidal physical models. Specific facilities include oscillating water tunnels and U-tubes, oscillating trays (including in-current flumes), the Total Environment Simulator and rotating facilities (i.e. race-track flumes, annular cells, Coriolis facilities). However, it should be noted that some facilities which come under the heading of 'standard' are in fact rare or unusual because either due to their large size (overcoming scaling difficulties), or due to additional special features. The HYDRALAB "Inventory of experimental facilities and instruments in Europe", (see www.hydralab.eu) provides an overview of such facilities, their sizes and characteristics. Some of the main types of facility are detailed below.

1.5.1   Current flumes

A current flume is a long and straight channel, with a free surface and pumps to generate steady or slowly-varying currents. Current flumes for river studies may have their main axis (x, in the direction of flow) inclined at a small angle to the horizontal. This angle can often be adjusted by a few degrees. Typical dimensions for current flumes are:

- Width = 0.3 m to 4.0 m
- Water depth = 0.1 m to 1 m
- Length = 10 m to 150 m

A current flume will generally have some or all of the following features:

- Pump(s) for generating a current. The pump(s) are likely to be centrifugal or axial flow. The choice will depend on the head loss in the system (axial flow pumps have a low head), the cost (centrifugal pumps are likely to be cheaper), the available sump capacity (axial flow pumps may require a lower sump capacity), the anticipated sediment load the pump must carry and the requirements for steady/ varying uni-directional/bi-directional flows;
- Current inflow system. The flow may be introduced into the current flume via a sump with an undershot or overshot weir to help even out the velocity profile

across the width of the flume. Alternatively the flow may be introduced via an enclosed expansion structure, which is designed to even out the flow distribution across the flume;

- Sediment recirculation system, or sediment feeder. In tests where sediment transport is expected to be significant, the mobile sediment bed may lower significantly during the test, unless sediment is fed into the system at the upstream end;
- Flow stabilisation zone. The flow emerging from the input system will have turbulence and will not have a natural boundary layer. The turbulence will dissipate and the current boundary layer will be generated as the flow progresses along the flume. A typical distance to allow for these processes to occur is a length of 10 to 20 times the water depth. The bed roughness of the flow stabilisation zone should be chosen to generate a similar bed shear stress to that expected over the test section;
- Mobile sediment bed, recessed into the floor of the current flume. The facility should be sufficiently long that the central test section is free of effects from inflow and outflow processes. It should also be sufficiently deep that the solid base of the test section does not affect the vertical distribution of pore-pressures in the bed;
- Sediment trap(s). One or more sediment traps are often built into the base of the flume to collect bedload sediment transport and may form an integral part of a sediment recirculation system;
- Outflow system. This may be a mirror image of the inflow system, particularly if the flume can be run in either direction.

The flow of a steady current along a flume can lead to the creation of secondary recirculation cells – slow but steady currents in the y, z plane. There will also be some side wall effects on the flow. The magnitude of these effects can be reduced to a generally insignificant level by keeping the ratio of current flume width, b, to water depth, h, in the range $b/h = 3$ to 5. This will minimise secondary circulation ($b/h > 3$) and avoid the development of non-uniform bed conditions ($b/h < 5$) (Whitehouse & Chesher, 1994).

If a structure is present (e.g. for scour studies), the blockage of the flow must also be considered. This might introduce a spurious laboratory effect into the observations due to the constraining effect of the flume walls causing flow to accelerate around the structure. It is generally considered acceptable if the cross-sectional area of an isolated structure (e.g. bridge piers) is less than about 1/6 of the total flume cross-sectional area.

1.5.2   Wave flumes

A wave flume is an open channel with a wave maker at one end (see HYDRALAB, 2007) and, in some cases, pumps to generate a collinear current. Wave flumes in Europe are listed in the HYDRALAB inventory of facilities at http://www.hydralab.eu. Ranges of dimensions available are:

- Width = 0.3 m to 5.0 m;
- Water depth = 0.1 m to 6 m;
- Length = 10 m to 324 m.

Figure 1.1 Wave flume at UPC.

An example of a wave paddle in a wave flume is shown in Figure 1.1. Wave flumes are used for sediment transport studies in two main ways: either sediment transport over a flat seabed, or on a beach. In cases where the study uses a flat seabed, this may be recessed into the base of the flume, or there may be a smooth, normally impermeable, approach slope (of ideally ≤ 1:20 and never steeper than 1:10) generally followed by a flat solid section before the mobile bed. The flat section should generally be at least one wavelength long to allow the waves to assume a steady form as they propagate over the mobile sediment bed. A passive wave absorber (normally a beach or permeable foam/mesh) dissipates wave energy that has propagated past the flat test section. Occasionally an active absorption system, consisting of a wave paddle that responds to incident waves, is used.

In flume tests where the evolution of a beach is studied, the bathymetry is moulded to the desired cross-shore profile. In many cases the offshore part of the bathymetry consists of a smooth, normally impermeable, approach slope (of ideally ≤ 1:20 and never steeper than 1:10 – as for the flat bed case) generally followed by a solid section representing the natural or representative beach slope, which should be at least one peak wavelength long. Wave flumes may be used to model the sediment transport around structures, provided that the blockage effect caused by the structure is not significant. Additionally wave flumes are used for testing the stability of structures, the influence of waves on vegetation and many more topics related to the behaviour of structures.

1.5.3 Wave basins

Wave basins are among one of the most common hydraulic research facilities at universities, research centres and commercial hydraulic institutions. Essentially, they are water tanks in which a wave field can be modelled in two horizontal dimensions. Typical dimensions of these facilities are about 20 m by 30 m and with variable water depths up to about 1 m. The largest basin in Europe has dimensions of 50 by 90 m. The dimensions of wave basins are chosen for modelling phenomena at scales between 1:10 and 1:100. The water depth is typically determined based on the primary applications of the basin. For offshore applications model facilities are designed for deep water conditions with 3 to 10 m depth so that water depths of several hundred meters can be reproduced.

Wave basins are equipped with wave generators – typically fixed on one side or consisting of moveable elements which can be placed for testing at different wave directions relative to the structure or coast subject to testing. Some wave basins are equipped with wave generators which can create a unidirectional wave field (2D, long-crested), but the wave generators can be moved such that various wave directions can be reproduced.

In recent years a number of wave basins have been constructed with multi-directional wave generators, which are able to produce waves at varying degree of complexity. These wave generators are normally placed along one side of the basin and the waves can then be produced at different directions relative to the basin walls. The wave generators can create a three-dimensional (3D) wave structure, also known as short-crested waves. An example is described by Aage & Sand (1984).

In order to reduce the un-warranted reflections from the basin sides and the structure subject to testing, passive absorbing side walls are installed in the basin either in the form of gently sloped beaches of crushed stone or absorbing elements consisting of net meshes or porous wooden or steel slopes. The most advanced wave generators are equipped with control systems to actively absorb wave energy reflected back towards them from model structures and basin walls.

1.5.4   Wave and current basins

A number of problems require testing in combinations of waves and current. Thus some wave basins are also equipped with current generation facilities. These can either be internal, which means that current flow is forced to circulate inside the basin or they can have external recirculation channels from where the water is distributed over (parts of) the basin boundaries. A third option is to circulate the water through a conduit under the basin floor.

The simplest conditions are when the current is perpendicular to the wave field (along the front of wave generators which is often the requirement for coastal models). Co-linear waves and currents can also be produced in some facilities, although current inflow through the bottom in front of the wave generator may give rise to unacceptable modification of the wave field.

Wave basins can be used to perform the investigation of wave/current interaction over a fixed or an erodible bed. Whereas in wave flumes, the current is co-linear with respect to the wave direction of propagation (e.g. Kemp & Simons, 1992), within wave basins the case of current crossing the wave field at a right-angle (or other specified angle) can be performed (Visser, 1986; Arnskov et al., 1993; Simons et al., 1995 [Figure 1.2], Andersen & Faraci, 2003). Special types of wave flume also exist, where waves interact with the current orthogonally, for example the oscillating tray of Sleath (1990), or the apparatus of the University of Catania (Figure 1.3) used by Musumeci et al. (2006).

Figure 1.2 Wave basin at Deltares.

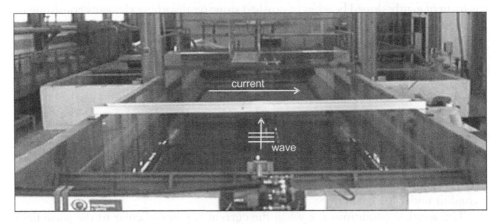

Figure 1.3 Experimental facility at the University of Catania used for the investigating the interaction of waves and current interacting at a right angle. In the interaction region a sandy bottom is present, in order to investigate bed morphodynamics.

A combined wave and current basin may give rise to various problems if the wave boundary (wave generator and/or guide walls) obstructs the flow. Typically more space is required in the basin if both current and waves are to be correctly reproduced, and even in large facilities it may be impossible to control wave and current generation to match prototype conditions. Therefore wave and current tests are often performed with compromise conditions.

Both in flumes and in basins, the current is usually generated by forcing a constant flow within the wave tank, by means of a pumping system. Great care must be used in the design of both the inlet and the outlet section of the current, since spurious effects, such as strong and unwanted secondary circulation or vortex shedding at the edge of the jet entering the tank, may strongly affect the flow in the area where the current and the waves interact with each other.

For mobile bed tests an issue which must be considered is how to ensure the continuous availability of sediments during the experiments. In particular, no sediments should be allowed to accumulate at the inlet and/or at the outlet of the current. In some facilities this is achieved by installing sediment traps and recirculating the sediment either manually or automatically.

An example of a specialised wave and current flume is the Total Environment Simulator (University of Hull). It is designed to simulate the combined effect of waves, currents and rainfall on sediments, which can be either cohesive or non-cohesive and animals and plants. The effective length of the flume is 11 m with additional inlet and outlet tanks to service the flow and sediment recirculation system and the depth is 1.6 m. One of the key advantages of the flume is the flexibility of the modelling space.

Importantly, this is one of the few facilities designed to operate with saline water and has day light lighting to enable plant growth. The flume can be used to investigate problems where one or more of these flow driving mechanisms need to be modelled to understand sediment transport and ecological problems. The working area of the flume can also be subdivided to allow channels of different dimensions to be investigated providing flexible control of the width to depth ratio and enabling parallel experiments to take place simultaneously.

1.5.5   Oscillating water tunnels

Oscillating water tunnels (OWTs) are closed ducts that have been developed to study details of processes in the wave boundary layer. An OWT can be described as a box of water which can be forced to move horizontally over a fixed or mobile bed. In this type of facility the particle oscillations under waves can be reproduced at a scale of 1:1, eliminating many of the problems of scaling. These facilities are ideal for detailed study of the hydrodynamic structure of the wave boundary layer and of water-seabed interaction under waves. OWTs often have the shape of a U-tube with a horizontal measuring section and a piston in one of the legs of the tube for the generation of the desired oscillatory flow. The dimensions of the tunnel should ideally be such that the phenomena under study are not affected by the boundaries. The measuring section generally has a length of 5–10 m, with a height and width of 0.5–1.0 m.

Some tunnels also have the facility to superimpose a current on the oscillatory flows. Figure 1.4 shows the Large Oscillating Water Tunnel of Deltares in the Netherlands (Ribberink & Al-Salem, 1994; Ribberink, 1989).

An important advantage of OWTs is that moderate and extreme wave-induced flows can be simulated under field-scale velocity conditions, so including the large Reynolds-number ranges typical of in the field. In the largest tunnels with test sections of 10–15 m length the more extreme wave conditions can also be simulated under field-scale wave-period conditions. Downscaling is not necessary and scaling

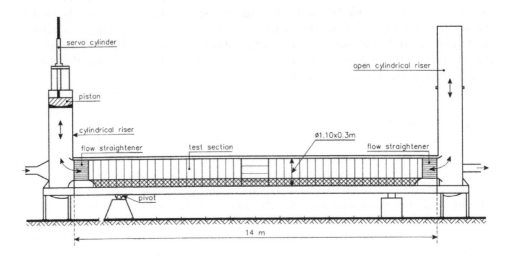

servo cylinder

open cylindrical riser

piston

cylindrical riser

Ø1.10x0.3m

flow straightener   test section   flow straightener

pivot

14 m

Figure 1.4 Large Oscillating Water Tunnel of Deltares (LOWT).

Table 1.2 Oscillating Water Tunnels.

Tunnel	Research institute	Waves/Current	Test section length (m)	Reference
U-tube	Technical University of Denmark	Waves	10	Staub et al. (1983)
Large Oscillating Water Tunnel (LOWT)	Deltares (Netherlands)	Waves and Currents	12	Ribberink & Al-Salem (1994), Ribberink (1989)
Aberdeen Oscillating Flow Tunnel (AOFT)	University of Aberdeen (Scotland)	Waves	10	O'Donoghue & Clubb (2001)
Water tunnel	Tokyo University (Japan)	Waves and Currents	4	Dibajnia & Watanabe (1996)

problems can therefore be avoided. Disadvantages of OWTs are the limitation to non-breaking waves and co-linear waves + current. It should also be realized that residual flows (averaged over the wave period) in OWTs are different than under real surface waves, due to the uniformity of the flow (no phase differences) and the absence of vertical orbital flows and a free surface.

In Table 1.2, an overview is given of a number of existing OWTs. The generation of the desired (regular or random) waves is mostly achieved with an electronically controlled piston (hard system) or with a pneumatic (soft) system (U-tube).

During mobile bed experiments with residual transport, a scour hole will develop at the upstream end of the test section, and since (the existing) tunnels have no sand feeding arrangement, this erosion will propagate into the test section, which limits the possible duration of an experimental run. Refilling with sand and restoring of the sand bed is then needed.

The test section is generally provided with (thick) glass side windows, which enables visual observation of the near bed processes and the use of laser-light and camera measuring techniques (Laser Doppler Anemometry, Particle Image Velocimetry). The roof of the test section may consist of a series of lids which can be opened individually for sand refilling and for instrument installation.

1.5.6   Annular cells

Annular cells are designed to investigate small scale morphodynamic processes induced by wave motion. A annular cell is made up of two concentric cylinders which are integral with two waterproof covers, the annulus between the two cylinders being filled with a thick layer of sand at the bottom and water up to the top. The cylinders oscillate together around the central axis, driven by a crank linked to a system of cogwheels and an electrical motor, remotely controlled by a PC. The system allows the motion of a shallow water cylindrical sinusoidal monochromatic wave to be reproduced, provided that the distance between the external cylinder and the internal one is small enough that the hypothesis of 2D waves is satisfied. The walls are transparent,

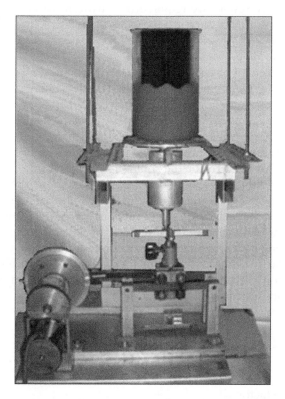

Figure 1.5 Side view of the annular cell used at the University of Catania by Faraci & Foti (2006). The bottom of the cylinder is filled with natural quartz sand, and the upper part is filled with water.

so that the flow field can be observed and recorded optically. Compared to more traditional facilities such as wave flumes or U-tube, the great advantage of the annular cell is related to the reduced dimensions of the apparatus (external diameter in the range 70–170 mm), which make it easy to set the control parameters in a very accurate way and also to repeat the same experiment a great number of times, so that statistical analysis can be performed.

1.6   GENERAL PRINCIPLES OF SCALE MODELLING

A physical model is in principle an analogue computer, where a physical parameter measured in the model represents the same parameter at the corresponding location in the prototype. In the case of scale models the quantity measured in the model has to be transformed by a scaling or model law to obtain the estimated magnitude of the actual prototype parameter.

A vital first step in the planning of laboratory experiments and physical modelling is consideration of the scaling laws that apply. This is essential for deciding the correct way to scale model results so that they can be interpreted quantitatively and at prototype scale (e.g. sediment transport rates). However, it is also important for cases where only qualitative results are required (e.g. patterns of erosion and deposition of sediment) as it is still necessary to reproduce the relative strengths of the forcing factors correctly. Neglect of scaling considerations can render model results either meaningless or misleading.

The general concept of physical modelling, the principles and criteria of similitude as well as the practical requirements and scaling laws to achieve similitude in hydraulic engineering models are well-addressed in many excellent textbooks (e.g. Langhaar, 1951; Hughes, 1993a). In hydraulic engineering, physical models have traditionally been defined as a simplified, *smaller* reproduction of objects, states and flow processes of their full-scale prototype counter parts (hereafter simply called 'prototype' or 'nature') which are of concern for the solution of problems in fluid mechanics and hydraulic engineering (biophysical models are sometimes larger than the prototype). To yield qualitatively and quantitatively useful results, physical model have to be geometrically, kinematically and dynamically similar to prototype conditions, dynamic similarity being the most important, followed by kinematic similarity. Although generally not recommended, geometrically distorted models are used in some cases. Full dynamic similarity between two geometrically and kinematically similar systems results from Newton's second law and requires that the ratios of all corresponding force vectors in both systems are equal:

$$\frac{(F_i)_n}{(F_i)_m} = \frac{(F_g)_n}{(F_g)_m} = \frac{(F_\mu)_n}{(F_\mu)_m} = \frac{(F_e)_n}{(F_e)_m} = \frac{(F_p)_n}{(F_p)_m} = \frac{(F_s)_n}{(F_s)_m} \qquad (1.1)$$

or in terms of scale factors:

$$N_{Fi} = N_{Fg} = N_\mu = N_{Fe} = N_{Fp} = N_{Fs} \qquad (1.2)$$

where F_i is the inertia force, F_g the gravity force, F_μ the fluid friction force, F_e the elastic compression force, F_p the pressure force and F_s the surface tension force, and indices n and m are for full-scale in nature and model scale, respectively. It can be demonstrated that full dynamic similarity is only possible at scale 1:1; i.e. Equation 1.1 and 1.2 cannot be fulfilled in any scaled model. Moreover, a partial dynamic similarity involving inertia forces and more than one governing force from the other five forces is also not possible, provided the model is not operated with a different fluid from prototype or/and another acceleration field than gravity (e.g. centrifuge). As physical models are smaller (or larger in case of biophysical models), the most crucial step in physical modelling is to find the most appropriate governing force which together with the inertia force should be scaled similarly while all other forces can be neglected in the model. Hence, the appropriate similitude law is obtained by equating the scale factor of inertia force N_{Fi} with one of the scale factors of the other forces in Equation 1.1. For instance, $N_{Fi} = N_{Fg}$ means that gravity force dominates, thus leading to the Froude similitude law, i.e. the Froude number must be the same in model and in nature $(Fr)_m = (Fr)_n$ with $Fr = v/(g \cdot h)^{1/2}$ (v and h: characteristic flow velocity and water depth, respectively). Moreover, $N_{Fi} = N_{F\mu}$ means that viscous friction forces dominate, thus leading to the Reynolds similitude law, the Reynolds number must be the same in model and prototype $(Re)_m = (Re)_n$ with $Re = v \, l/\mu$ (v, l: characteristic flow velocity and length). The scale factors for various physical quantities resulting from Froude and Reynolds scaling laws are compared in Table 1.3.

When applying for instance a Froude similitude, gravity forces are properly scaled while the other forces are subject to the so called *scale effects*, i.e. they are overestimated in the model when Froude scaling to prototype conditions is applied. The degree of overestimation strongly increases as the model scale becomes smaller, thus illustrating the advantage of large testing facilities. The scale effects arise, because dynamic similarity cannot be fulfilled simultaneously in the same model for the other forces. Some of the numbers used as bases for similarity similar similarity numbers used in hydraulic engineering are given in Table 1.4, including the definition of parameters and dimensions in Table 1.5.

1.7 LAYOUT OF THE BOOK

This book covers five cross-disciplinary themes in laboratory physical modelling, waves, breakwaters, sediments and ecology. The contents of each, for which the respective lead authors are responsible, are briefly described below. It is the intention of the authors that the separate chapters can either be used as standalone guidelines to particular types of experimentation or together as a reference document. It reflects the combined knowledge and experience of the researchers from the major laboratories across Europe and is therefore unique in both its content and scope.

1.7.1 Wave modelling

This chapter describes experimental reproduction of wind, waves and swell. It focuses on applications for the modelling of coastal processes and for the support of planning

Table 1.3 Scaling factors resulting from Froude and Reynolds similitude laws.

Notation	Unit		Froude	Reynolds
Length, width, height	m	l	N_l	N_l
Surface	m²	l²	N_l^2	N_l^2
Volume	m³	l³	N_l^3	N_l^3
Time	s	t	$N_l^{1/2}$	$N_l l^2$
Velocity	m/s	lt^{-1}	$N_l^{1/2}$	$N_l l^{-1}$
Acceleration	m/s²	lt^{-2}	I	$N_l l^{-3}$
Discharge	m³/s	$l^3 t^{-1}$	$N_l^{5/2}$	$N_l l$
Kinem. viscosity	m²/s	$l^2 t^{-1}$	$N_l l^{3/2}$	I
Mass	kg	m	N_l^3	N_l^3
Force	N = kgm/s²	mlt^{-2}	N_l^3	I
Density	kg/m³	ml^{-3}	I	I
Specific weight	N/m³	$ml^{-2}t^{-2}$	I	N_l^{-3}
Dyn. Viscosity	kg/ms	$ml^{-1}t^{-1}$	$N_l^{3/2}$	I
Surface tension	kg/s²	mt^{-2}	N_l^2	N_l^{-1}
Elasticity	kg/s²m	$ml^{-1}t^{-2}$	N_l	N_l^{-2}
Pressure and tension	kg/ms²	$ml^{-1}t^{-2}$	N_l	N_l^{-2}
Momentum (impulse)	kg m/s	mlt^{-1}	$N_l^{7/2}$	N_l^2
Energy	kg m²/s²	$ml^2 t^{-2}$	N_l^4	N_l
Power	kg m²/s³	$ml^2 t^{-3}$	$N_l^{7/2}$	N_l^{-1}

and design of marine structures to be constructed in the real world. It includes deep and shallow water, near-shore and offshore problems and all types of fixed and float-ing structures. By describing the short-comings and concerns of some experimental practices, this chapter is expected to serve as inspiration for the academic commu-nity. Although not specifically treated therein, experiments with other types of waves such as ship waves, waves caused by land-slides and tsunamis, may benefit from the methods described in this chapter.

1.7.2 Breakwaters

This chapter deals with the physical modelling of fixed-bed, rubble mound breakwater structures under wave influence. It covers best practice with regards to model set-up and operation, analysis and reporting procedures whilst identifying areas for future work. The modelling of other breakwaters and coastal structures are not discussed in detail. It therefore excludes impermeable structures, vertical walls and composite structures. It also deals only with short-term, short-wave models and excludes the effect of currents. Furthermore, moveable-bed models and structure-foundation interactions are not discussed.

1.7.3 Sediments

This chapter synthesises and documents procedure and practice in the laboratory physi-cal modelling of sediment transport in rivers, estuaries and the sea. It does not attempt to

Table 1.4 Similarity numbers applied in fluid mechanics parameters definitions (supplemented and modified from White, 1994).

Parameter	Definition	Qualitative ratio of effects	Importance
Reynolds number	$Re = \dfrac{\rho UL}{\mu}$	$\dfrac{\text{Inertia}}{\text{Viscosity}}$	Always
Mach number	$Ma = \dfrac{U}{c}$	$\dfrac{\text{Flow speed}}{\text{Sound speed}}$	Compressible flow
Froude number	$Fr = \dfrac{U}{\sqrt{gL}}$	$\dfrac{\text{Inertia}}{\text{Gravity}}$	Free-surface flow
Weber number	$We = \dfrac{\rho U^2 L}{Y}$	$\dfrac{\text{Inertia}}{\text{Surface tension}}$	Free-surface flow
Cavitation number (Euler number)	$Ca = \dfrac{P - P_v}{\rho U^2}$	$\dfrac{\text{Pressure}}{\text{Inertia}}$	Cavitation
Prandtl number	$Pr = \dfrac{\mu c_p}{k}$	$\dfrac{\text{Dissipation}}{\text{Conduction}}$	Heat convection
Eckert number	$Ec = \dfrac{U^2}{c_p T_0}$	$\dfrac{\text{Kinetic energy}}{\text{Enthalpy}}$	Dissipation
Specific-heat ratio	$k = \dfrac{c_p}{c_v}$	$\dfrac{\text{Enthalpy}}{\text{Internal energy}}$	Compressible flow
Strouhal number	$St = \dfrac{\omega L}{U}$	$\dfrac{\text{Oscillation}}{\text{Mean speed}}$	Oscillating flow
Roughness ratio	$\dfrac{\varepsilon}{L}$	$\dfrac{\text{Wall roughness}}{\text{Body length}}$	Turbulent, rough walls
Grashof number	$Gr = \dfrac{\beta \Delta T g L^3 \rho^2}{\mu^2}$	$\dfrac{\text{Buoyancy}}{\text{Viscosity}}$	Natural convection
Temperature ratio	$\dfrac{T_w}{T_0}$	$\dfrac{\text{Wall temperature}}{\text{Steam temperature}}$	Heat transfer
Pressure coefficient	$C_p = \dfrac{P - P_A}{\frac{1}{2}\rho U^2}$	$\dfrac{\text{Static pressure}}{\text{Dynamic pressure}}$	Aerodynamics, hydrodynamics
Lift coefficient	$C_L = \dfrac{L}{\frac{1}{2}\rho U^2 A}$	$\dfrac{\text{Lift force}}{\text{Dynamic force}}$	Aerodynamics, hydrodynamics
Drag coefficient	$C_D = \dfrac{D}{\frac{1}{2}\rho U^2}$	$\dfrac{\text{Drag force}}{\text{Dynamic force}}$	Aerodynamics, hydrodynamics
Peclet number	$Pe = \dfrac{VL}{D} = Re * Sc$	$\dfrac{\text{Advection}}{\text{Mass diffusity}}$	Turbulence, suspensions
Schmidt number	$Sc = \dfrac{\mu}{D}$	$\dfrac{\text{Mass diffusity}}{\text{Momentum diffusity}}$	Turbulence

Table 1.5 Parameter definitions (supplemented and modified from White, 1994).

Quantity	Symbol	Dimensions MLT θ
Length	L	L
Area	A	L^2
Volume	V	L^3
Velocity	υ	LT^{-1}
Acceleration	dV/dt	LT^{-2}
Speed of sound	c	LT^{-1}
Volume flow	Q	L^3T^{-1}
Mass flow	m	MT^{-1}
Pressure, stress	p, σ	$ML^{-1}T^{-2}$
Strain rate	ε	T^{-1}
Angle	θ	None
Angular frequency	$\omega = \dfrac{2\pi}{T}$	T^{-1}
Oscillation period	T	T
Viscosity (dynamic)	μ	$ML^{-1}T^{-1}$
Kinematic viscosity	ν	L^2T^{-1}
Surface tension	Y	MT^{-2}
Force	F	MLT^{-2}
Moment, torque	M	ML^2T^{-2}
Power	P	ML^2T^{-3}
Work, energy	W, E	ML^2T^{-2}
Density	ρ	ML^{-3}
Temperature	T'	θ
Specific heat	c_p, c_υ	$L^2T^{-2}\theta^{-1}$
Specific weight	γ	$ML^{-2}T^{-2}$
Thermal conductivity	k	$MLT^{-3}\theta^{-1}$
Expansion coefficient	β	θ^{-1}
Mass diffusion coefficient	D	L^2T^{-1}

M = Mass, L = Length, T = Time, θ = Temperature

cover all aspects of physical modelling but draws together aspects in which the partners involved have a special expertise, and hence it complements the standard texts. It encompasses both laboratory experiments to elucidate process knowledge (research) and physical modelling of site-specific applications (consultancy). Its remit includes most types of physical modelling of sediments, rather than focussing in depth on one type of application. It excludes soil mechanics and geotechnics, fluidisation and permeable flows.

1.7.4 Ecology

Physical modelling of ecology is essential to improve our understanding of both the impact of environmental factors on biota and the impact of biota on their environment. Research on the interaction amongst water flow, morphology, sediment transport and biological processes is essential for better understanding of the natural environment and to improve management of the natural environment. This is a very new area of research and the challenges associated with it are many. Added to the standard problems of hydraulic experimentation are issues of plant and animal health and (for animals) ethical considerations. This chapter covers current best practice for incorporating plants/animals into hydraulic studies. It discusses issues associated with experimentation with both living organisms and inert surrogates.

1.7.5 Composite modelling

This chapter describes techniques for combining physical and numerical modelling in order to improve the physical modelling infrastructure for coastal systems. Composite modelling can lead to different forms of improvements: being able to model problems that cannot be modelled by either physical or numerical modelling alone; increasing quality at the same cost or obtaining the same quality at reduced cost; reducing uncertainty at the same cost and realising that uncertainty reduction is also a quality issue.

Since composite modelling in the field of coastal modelling is a rather new approach, this chapter focuses on outlines and summaries of the seven composite modelling study cases that were conducted in the HYDRALAB CoMIBBS project. Although not specifically treated herein, composite modelling of other near-shore processes may benefit from the methodologies and techniques described in this section.

Chapter 2

Waves

Jens Kirkegaard

Agustin Sanchez Arcilla, Mark Klein Breteler, Rui Capitão,
Neelke Doorn, Juana Fortes, Hemming Schäffer,
Reinold Schmitt-Koppenhagen, Carl-Trygve Stansberg &
James Sutherland

2.1 INTRODUCTION

Understanding the impact of water waves is essential for design of marine structures and assessment of coastal development due to natural causes as well as human intervention. The use of physical and mathematical models is common as part of design processes and impact assessments. Although numerical models have replaced physical models for a number of problems, particularly project models covering large areas, there are still a large range of problems that today are more adequately treated in physical models. On the other hand, numerical models also serve to improve wave modelling, particularly with respect to definition of boundary conditions for wave tests.

Consequently, there is a strong need for developing systematic methods for interaction of numerical and physical models, which may be expected to be complementary tools for many years to come. Models will always represent the full-scale problem with some simplification. Thus, the modeller is forced to compromise on the details of the model.

Realistic answers to wave impact problems can only be obtained if the area represented in the model contains all important characteristics of the wave field that are relevant for the problem in question. This is the reason why researchers have improved the wave reproduction methods for laboratory use. A number of problems will require further scientific development – such as those presented later in this report: representation of long waves in models and multi-directional wave generation of natural sea states.

2.2 APPLICATION OF WAVE MODELS

The purpose of wave testing and thus the applications of wave models fall in the following categories:

- Wave interaction with fixed structure (breakwaters, seawalls and jetties, bridge piers, tanks, platforms, pipelines, intakes and outfalls). For practical aspects related to rubble mound breakwater modelling refer to chapter 3;
- Wave interaction with floating structures (ships, pontoons and caissons, buoys, floating equipment and machinery and marine construction processes);

- Wave interaction with the seabed and coasts (movable bed models of sediment transport and erosion);
- Wave kinematics (freak waves, wave transformation – dissipation);
- Wave-generated currents (impact on dispersion processes in the sea and on sediment transport).

The term interaction refers to the impact on the structure (pressure and forces), the transformation of the wave caused by the structures (for example diffraction, run-up and overtopping) and response of the structure (for example displacement of breakwater armour blocks, ship motions and mooring forces).

Typical applications of experimental wave testing are:

- Coastal development – morphology, cross-sections, beach and dune profile;
- Scour (local erosion);
- Impact of coastal structures;
- Wave penetration in harbours (wave agitation, disturbance);
- Behaviour of ships and offshore structures in waves;
- Wave (inter)action with structures (stability, overtopping, run-up on structures);
- Impact due to green water and negative air-gap on offshore structures;
- Ship waves;
- Waves on current;
- Wave energy devices.

The most common facilities used for experimental wave testing include wave flumes, wave basins, multi-directional wave basins and deep water wave basins (refer to chapter 1).

2.3   SELECTION OF WAVE CHARACTERISTICS FOR MODEL TESTS

2.3.1   Types of waves in hydraulic models

Wave modelling – or reproduction of waves – includes waves of different complexity as defined below:

- Regular waves (mono-chromatic waves, sinusoidal waves);
- Irregular waves (random waves, natural sea waves);
- Multi-directional waves;
- Other (transient waves, bi-chromatic waves, etc).

The selection of wave conditions in the model must be carefully considered in relation to the actual problem that needs a solution. In the following sections, the important considerations for selection will be described using representative wave time series, uni-directional/ multi-directional waves and the importance of long waves.

Figure 2.1 Impact of detached breakwater tested in coastal model facility.

2.3.2 Model scale ratios

Generally, the largest possible scale is selected for the available test facility. Thus, the size of the laboratory basin or flume is a primary parameter. There are many reasons for this including consideration of model laws other than Froude scaling (Reynolds, Weber), measuring accuracy and scale selection related to problem under study (e.g. global or detailed impact). Other limitations than physical dimensions, may play an important role in scale ratio selection, for example, the wavemaker capacity, towing speed, etc.

Although this suggests that there is a wide range of possibilities when modelling according to Froude scaling, one has to accept that the scale selection should assure that the dominant forces are well represented. For wave reproduction, this means that capillary forces should play an insignificant role only. Consequently, the wave lengths must not be too short. A practical lower limit is 0.3 m corresponding to a wave period of 0.4 s. No essential part of the wave spectrum should be below this value.

Very small scales also affect the repeatability by which the small-scale waves can be reproduced in the laboratory. In tests where viscous forces are important, due attention has to be paid to the Reynolds number. This is related primarily to the structure being tested rather than to the ability to reproduce the waves.

2.3.3 Selection of representative sea states

Wave characteristics from the actual project location shall be the basis for selection of representative sea states for a model test programme. Often, for example in offshore engineering, the specification is determined by the project owner based on statistical analysis of wave conditions at the site. However, the experience and traditions of the actual laboratory must be considered for selecting the realistic laboratory wave conditions.

There are a number of problems related to the selection in practice:

- Wave recordings, if available, have often been made over relatively short time spans. Thus, they may not be sufficient to establish a credible statistics;
- Field wave recording is often not directional, which means that directions at best must be determined based on meteorological conditions;
- The individual time series are short, which is a problem particularly to determine the possible long wave parts of the spectrum;
- Some wave recorders cannot describe long waves due to systematic electronic filtering (accelerometer buoys) or they are unable to describe the high frequency part of the spectrum (pressure sensors in deep water);
- Wave recordings with double peaked spectra – often with different directions – are by some analysis procedures described simplistically by mean characteristics;
- Directional spread calculations often poor as only three signals are used in computation;
- The spectral shape of sea states varies and is often not available and documented for extreme sea states.

Some of these problems must be handled by use of numerical wave models. It is now possible to obtain long series of wave data (typically 15–20 years) – at least for deep water locations – generated from historical meteorological data. This information can then be transferred to nearshore locations by numerical models and compared with measured data for shorter time spans. By this method, it may be possible to derive parametric expressions for extreme wave conditions. It is important, however, to use estimated extreme waves with great care. Therefore it is advisable to use sensitivity tests with variations of spectral parameters.

The final selection of representative wave conditions should be made with due consideration of the test objective. The selection depends on the application (e.g. wave impact on caisson structures versus rubble mound breakwaters). Experience and skills of the study team are important factors when classifying the problems. If time and budgets are available, it is recommended to do sensitivity tests.

2.3.4   Duration of time series

Many projects – especially those involving dynamic structures – require very long testing time to obtain enough information to derive design values, i.e. values with a well-defined probability. Examples of such projects include motions of moored ships and platforms due to long natural oscillation periods, stability of rubble mound structures and wave forces on elevated decks.

The expected value of the maximum wave height in a wave time series is a function of the duration which conditions the length of the synthetically generated time series (since the expected value of the maximum must be generated by the paddle). Traditionally, 1–3 hour sea states (with the design (H_s, T_p) values) have been reproduced and either the maximum value or a value from a statistical analysis of all extremes (preferred from a statistical point of view) has been used as the design load/response. The shorter durations may be acceptable when the tidal exposure is short. However, it has often been realised that either a much longer simulation (6 hours, 12 hours)

Figure 2.2 Wind turbine foundation exposed to breaking wave.

or repeated simulations with different wave time series (but identical (H_s, T_p) values) may be required to provide reliable estimates of the load/response. This is both time consuming and expensive.

It should therefore be considered if relatively short duration tests can be designed to produce as much information as long duration tests. This may be achieved by focusing the content of the wave time series on the wave sequences that will produce the extreme responses using methods such as 'importance sampling' or use of pre-determined group.

There is a need for systematic methods to extract short subseries of wave records, which can provide test results with the same reliability as traditional long series. This involves assessment of the return period of wave conditions versus the return period of response. In relation to duration of tests series, it is important to consider the actual shape of a storm which may influence the response of a beach or a breakwater section.

The possible appearance of exceptional waves may also be of potential interest for testing particularly sensitive structures. Such waves may require artificial stimulation. The study of exceptional waves is still at the research stage, particularly in shallow water.

2.3.5 'Free' and 'bound' long waves

Any structure or coastline is highly reflective when exposed to long waves. While long waves can often escape to deep water under natural conditions, they are inevitably entrapped in traditional laboratory experiments. Active absorption systems (absorbing wavemakers) can assist the transparency of offshore boundaries to long waves. Although much of the long-wave activity is generated locally, some conditions call for a non-linear wave-generation procedure at the incident-wave boundary. The current limitations to understandings of long waves limits reduces the ability to accurately reproduce natural long-wave conditions in the laboratory, even with the use of second-order wave generation and active absorption.

Long waves in nature (periods of the order 0.5–5 minutes) result from several generating mechanisms including second-order interactions between primary waves and low-frequency breakpoint oscillations. Although the consequence of these generating mechanisms may nowadays be quantified when focusing on small coastal regions, there is no way to assess the presence of long waves generated at neighbouring or remote stretches of coastline. Numerical models cannot cover sufficiently large areas to answer this fundamental uncertainty.

The long-period waves are particularly important for study of the behaviour of floating structures in shallow water such as moored large tankers and container vessels. These moored systems have natural frequencies in the same range as the long waves and are therefore susceptible to resonant motions with potential overload of mooring lines and/or impact on cargo transfer. Long waves are less prominent in deeper water because the bound long waves there are smaller relative to the height of the storm waves and swell.

Long wave problems are appreciated by laboratories working with shallow water hydraulics as important and there is generally a desire to take long wave problems into account in model test planning, design and execution. Several laboratories have developed wave generator control and active absorption by which they are able to take these effects into account (to a certain degree). Several research papers on the subject have been published over the last 25 years (refer to bibliography section on longwaves).

Figure 2.3 Three-dimensional wave during North Sea storm.

Although solved in principle, there are serious practical problems in handling of long waves by the wave generator. In shallow water, the necessary stroke of the wave generator is very large and may therefore restrict reproduction of extreme sea states at a satisfactory model scale.

The different extent of experience by the various laboratories gives rise to research needs ranging from development of control and absorption technologies to integration of numerical and physical models (composite modelling). There is still a need for more information from the nature and other sources, of the statistics of occurrence and phenomenological descriptions. Numerical models can be used to analyse trapping and possible undesirable trapping of long wave energy in scale models.

2.3.6   2D/3D waves

Multi-directional wave testing facilities have been available for several years now and experience has been gained on the benefits resulting from these more natural waves relative to traditional long-crested waves. Whereas the benefits of using irregular waves instead of regular waves were quickly accepted by the laboratories and users, a similar acceptance is still lacking for 3D waves. This is most likely due to a general experience that 2D testing is sufficient (or gives a conservative solution) for a large range of problems. The general lack of good quality directional wave data is also used as a reason not to use 3D testing. Newer field equipment (such as Acoustic Doppler Profilers and Directional Buoys) will eventually eliminate that concern.

The research in wave basins of the HYDRALAB-partners may be classified into the three main research fields – coastal engineering, offshore engineering and naval architecture.

In the field of coastal engineering, physical modelling in wave basins is used to account for the problem of wave disturbances in harbours, scour protection, wave transmission over submerged breakwaters, wave load on and diffraction behind breakwaters, overtopping at vertical and sloped structures as well as wave run-up at dikes. The effect of 3D waves compared to 2D waves was so far analysed in depth for the process of oblique wave run-up and wave overtopping at sea dikes (Oumeraci et al., 2001) in the wave basins of Franzius-Institute, Germany, and NRC, Canada. The results revealed no significant influence of multi-directionality (Möller et al., 2001, Ohle et al., 2002, Ohle et al., 2003). However, large effects of 3D waves are to be expected when analysing problems of diffraction (Daemrich, 1996), combined diffraction and transmission (Eggert et al., 1982) as well as wave loads on structures (Hiraishi, 1997, Franco et al., 1996).

In the field of offshore engineering, investigations in physical modelling in wave basins are related to questions on wave load. Comparative studies with 2D and 3D waves were carried out for moored semi-submersibles in the wave basin of MARINTEK, Norway, (Stansberg, 1997). These tests revealed that whilst linear in-line wave forces on semi-submersibles are reduced only by approximately 10–15% in multi-directional waves, the corresponding in-line slow-drift forces are overestimated up to 40% when using unidirectional waves instead of multi-directional waves. This stresses the need for further investigations to avoid designs of offshore structures that are too conservative. Transversal loads are, as expected, clearly increased in multi-directional waves.

In the field of naval architecture, physical wave modelling focuses on ship stability and seakeeping. Both 2D and 3D waves are used within the standard 'ship design by analysis' (Hirayama, 1997).

Problems in 3D wave modelling

The dominance of 2D wave modelling results from several complications and unsolved problems of 3D wave modelling. The description of representative sea state at model boundaries can cause issues for modelling. In 3D modelling, there is a need for more accurate sets of boundary conditions in order to bring out the advantages compared to 2D models. It is proposed that minimum standards of quality need to be defined for 2D and 3D modelling.

In 3D wave basins, the treatment of boundary effects such as wave reflection or damping at the boundaries of the wave basin, requires larger efforts than in 2D modelling. Although a lot of research has been carried out on this subject (e.g. Schäffer *et al.*, 2000), more research is needed in order to get better ratios of utilisable area of the basin, total surface area of the basin and absorption control.

Monitoring of 3D wave field within the basin can also be problematic. In 3D basins, wave conditions are more inhomogeneous in space than in 2D basins. Therefore, wave conditions in the 3D basins require higher resolution monitoring of wave parameters. Standard techniques such as resistance wave gauges, cannot provide an overall view of the wave field. New (or re-visited) measuring techniques, like stereo photogrammetry (Santel *et al.*, 2002), particle image velocimetry (Stagonas & Müller, 2007), stereo imaging (Wanek & Wu, 2006) or Starry Sky (Valembois, 1951) should be developed for use in physical wave modelling.

2.4   WAVE GENERATORS AND WAVE GENERATION

Since the start of wave modelling, many different wave generator systems have been developed. In 1951, Biésel & Suquet published their fundamental report on analytical solutions to wave generation principles for a number of different wavemaker types. The paper presented the transfer functions for the wavemaker paddle displacements to wave amplitude. The Biésel transfer functions are fundamental for wave generation.

2.4.1   Wave generator types

Today the most common wave generator types are:

- Piston (vertical paddle covering the full water depth or elevated above the bottom);
- Flap (hinged at or above bottom of tank);
- Wedge type (vertical paddle moving up and down a slope and suited for deep, intermediate and shallow water waves);
- Double-articulated, hinged;
- Snake type (2D/3D wavemaker), which may consist of one of the above types of wavemakers placed side by side.

Both piston and hinged types are used for unidirectional and multi-directional wave generators.

2.4.2   Actuators

All types of paddles used in wave generation are driven by one or more actuators. Hydraulic rams with a servo valve were the most commonly used before the 90s. Due to their excellent performance, they remain in use in a large number of laboratories. Recently, however, linear electric drives have been adopted in most new wave generators.

2.4.3   Control signals

The actuators work according to a control time series from a signal generator. The principles used in signal generators depend on the type of wave to be generated as listed below:

- Regular waves by sine generator;
- Synthesis based on specified standard or custom spectrum;
- Direct reproduction of measured wave time series;
- Directional distribution functions;
- First-order, second-order, cnoidal, solitary;
- Active control of wave groups;
- Active control of low frequency input.

The duration of signals was dependent on the capacity of the control computers, but with present days PCs there are no practical limitations on the duration. The probable maximum wave height increases with duration for a given mean energy of the wave signal. This should be considered carefully when selecting test duration for structures sensitive to high individual waves.

The wavemaker requires a gradual increase and decrease of the control signal before and after the test. The ramping up and down period depends on the laboratory size and is typically in the order of five seconds (for a 80 m long basin, 20 s is adequate).

When using active wave absorption control, the wave signal generator can be a separate computer-controlled system or it can be linked to the data acquisition system which is very convenient for subsequent analysis of test results. A system with active absorption control needs a very high speed of communication to perform an efficient and accurate compensation of the incoming wave series.

2.5   PLANNING AND EXECUTION OF TESTS

2.5.1   Bathymetric model construction

The main issues relating to construction of wave models are the reproduction of the bathymetry and the permeability and porosity of structures.

Correct reproduction of waves in a model is governed by the wave generator and by the bathymetry in the model. The most important issue is that the waves generated by the wavemaker and the water depth in front of the wavemaker are compatible. Large waves generated in too shallow water will break on the paddle and will not create realistic model waves.

In many cases, the horizontal dimensions of the wave basin or flume are such that waves cannot be generated in the correct water depth if the entire model is constructed to scale. The model therefore has to be 'truncated' and a steeper artificial slope must be constructed down to the basin floor. In such a case, there is a risk that waves are forced to break on the slope which may not be in agreement with the natural conditions. To avoid this, the modeller must use a combination of experience and analytical tools to find the optimal compromise between the model scale and the basin/flume dimensions.

The permeability and reflection characteristics of the model (typically for harbour and coastal models) must be reproduced in order to obtain realistic transmission of waves into the model. This aspect is covered in chapter 3.

2.5.2   Wave reflection in laboratory

Wave basin and flumes are confined water bodies and as such, there is a risk that the wave energy introduced through the wave generators is trapped and amplified in an unrealistic manner. The modeller thus needs to consider carefully how to measure and analyse the reflection and how to avoid unrealistic effects.

The calculated reflection coefficients are sensitive to the technique used and to the available time series recorded. In effect, depending on the number and type of series recorded and the technique used to split incident and reflected waves, this coefficient may show variations of a factor 10.

As in nature, waves are reflected from coastlines and structures. Ideally, a model area should be controlled actively on all boundaries. However, this is unrealistic for a number of reasons but primarily due to the complexity and costs. Common practice is to absorb the energy that reaches the basin boundaries by passive absorbers and to minimise the wave energy displaced outside the area of interest.

The total basin area is also an important factor with larger areas resulting in lower reflection problems. The minimisation of wave reflection is done by careful design of guide walls and passive absorbers. Guide walls must be placed in such a way that a minimal amount of energy is diffracted outside the area of interest, but without impacting on the wave field in the area of interest such as a port entrance. Passive absorbers should be designed to effectively reduce the wave energy. Gentle crushed stone absorbers or parabolic shaped 'beaches' are usual passive absorbers in the laboratory. Mesh baskets, multi-screen arrays & expanded foam are also used.

Some wave energy is reflected back – directly or indirectly from guide walls – towards the wave generator paddle. Absorption on the wave paddle is now in regular use in wave flumes. The advantage here is that the wave direction is well defined – opposite the original wave.

In wave basins, it is much more complex to absorb the reflected wave on the wave board. With a unidirectional wavemaker, the absorption can only relate to an average compensation along the wave board, but the reflected wave reaching the wave generator will often not be constant (in phase and amplitude) along the wave generator. A multi-directional wave generator is required to cope with this problem. In the absence of active absorption by the wave generator, the modeller has to compensate the specified control signal for the re-reflected wave energy so that the total wave energy generated and reflected by the paddle is in accordance with the required wave field. This may only be possible to a limited extent.

The principles described above relate to both short waves (sea and swell) and long waves. Long waves, which typically in models have a period of ten times the wind waves, are much more difficult to absorb. They require very wide – and in most cases unrealistic wide absorbers – and the active absorption can only be accomplished with large stroke wave generators.

In many cases, the only realistic option to minimise the unrealistic effect of long waves is to place the wave generator such that long wave energy is not trapped between the wave generator and the model. In some cases, it is practical to use a numerical model to determine the optimal wave generator position for the long wave frequencies found in the selected wave series.

2.5.3   Test programme

Detailed planning of the test programme is of paramount importance. The involvement of the user of the results is important, but the experience of the laboratory is essential to render credibility to the test results. Inadequate planning of a model study can result in systematic errors, whereby essential questions are not answered. This will be the case regardless of how much attention has been spent on selecting the optimal scale ratios. Such errors will be beyond scale and laboratory effects and may have equally important consequences.

In order to create an effective test programme it is recommended to:

- Include verification tests comparing wave input for conditions identical to the natural conditions;
- Include sensitivity tests, e.g. related to direction;
- Include a simple check on 'trivial' effects, e.g. heave transfer functions vs. theory (quality control);
- Use 3D waves if the results are very dependent on direction (approach channels to harbours, offshore moorings);
- For strongly non-linear phenomena such as green water and negative air-gap, a large number of sample spectrum realisations may be required to reveal a robust information on extreme statistics;
- Use hybrid modelling to expand the results of experiments.

Often, this may be regarded as an unnecessary expense; however, questions about the validity of the results will often be asked at a later stage – typically after the model does not exist anymore.

2.6 MEASUREMENT AND ANALYSIS
OF LABORATORY WAVES

2.6.1 Measurement

Waves in the laboratory must be documented in order to verify their agreement with sea states specified for the tests. The documentation comprises of wave conditions in front of wave generator and at reference points for verification and impact analysis.

The set-up of wave gauges should be planned such that erroneous information, such as from reflections and trapped energy, may be identified. The information required in its simplest form is the variation of water elevations at the point(s) of interest. Direction measurements (multi-directional waves or reflected waves) require either an array of position gauges or measurements of orbital velocity measurements in the water column.

Research on wave measuring technology

It appears that several laboratories are presently active in some kind of research on wave measurements. There is also a general request for more elaborate measuring techniques like video imaging techniques providing surface elevation time series in a dense grid (Santel *et al.*, 2002; Wanek & Wu, 2006; Stagonas & Müller, 2007). There may be a need for a joint research action to advance these techniques.

Wave measurements may be categorised in various ways. He present approach is to look at the physical quantity and the dimensions in which it is measured. As all measurement techniques can now provide results in discrete or continuous time, the categorisation does not consider the temporal variation. Nevertheless, possible sampling rates are of interest.

Wave measurements in basins and flumes are made to obtain information about the position of the fluid boundary and the particle velocity inside the fluid domain. The part of the fluid boundary relevant for wave measurements is the free surface elevation. The surface elevation is generally a 2D surface that may usually be described as a function of horizontal space and time, $\eta(x, y, t)$. The simple point measurement of surface elevation (0D) thus provides $\eta(x_0, y_0, t)$. Here, (x_0, y_0) is typically a fixed point, but it could also be moving such as with a towing carriage. Profile measurements (1D) provide e.g. $\eta(x, y_0, t)$ for some interval of x, while full surface measurements (2D) provide $\eta(x, y, t)$ over an area.

The velocity is generally a 3D vector function, where each of the three components varies in 3D space, $\mathbf{u}(x, y, z, t) = (u(x, y, z, t), v(x, y, z, t), w(x, y, z, t))$. Measurements may be uni-axial, bi-axial or tri-axial to provide one, two or three of the vector components. Furthermore, they may be point measurements (0D), profile measurements (1D), plane measurements (2D) or in principle also measurements over 3D space. The most basic velocity measurement is a uni-axial single point current meter providing e.g. $u(x_0, y_0, z_0, t)$. The most advanced technique one may think of would be perhaps some holographic technique providing all three velocity components in 3D space, $u(x, y, z, t)$.

2.6.2   Wave analysis

Analysis of short and long waves

For analysis and interpretation of test results, it is often required to separate the measurements in short and long waves. This is particularly important when wave agitation of floating bodies are under study, as these systems often have a range of different eigen frequencies. The separation can be either related to a fixed frequency or by detailed analysis of the wave spectrum. For practical reasons, it may be considered to separate at about 20 s wave period; however, in some cases other values may be more relevant. This will depend on the field location and on the actual type of marine structure.

Coping with reflecting waves

Reflected waves are unavoidable in wave models. The preferred strategy is to plan the model with a minimum of harmful reflection, but in any case it is necessary to document them. This may be done by an array of wave gauges, consisting of five gauges (linear for 2D or quasi 3D only). There may be a need for defining a standard for positioning of the gauge array relative to the structure, in relation to the wave length (0.5–1.0 L) to be tested and techniques for single group reflection tests.

The most common methods for resolving 2D spectra into incident and reflected components include the two probe, one phase angle method of Goda & Suzuki (1976) the three probe, two phase angle method of Mansard & Funke (1980), the three probe method of Isaacson (1991), the vertical array of probe plus velocity method of Guza et al. (1984) or Hughes (1993b), the co-located velocities method of Hughes (1993a), and Zelt & Skjelbreia's (1992) extension of the 3-probe method to an arbitrary number of wave gauges, which introduced weighting functions to try to minimise the effect of having probe spacings close to a multiple of half a wavelength. All the methods rely on the linear superposition of many wave components; no non-linear interactions are represented. Frigaard & Brorsen (1995) introduced a method to separate incident and reflected wave fields in real time using theoretical phase shifts and digital filtering – see also Baldock & Simmonds (1999).

Methods that use phase information perform better than methods that do not, while methods that use three (or more) probes can have a wider useful frequency range than 2-probe methods. These methods can all be extended to the quasi-2D case, where unidirectional waves are reflected from a structure at a known angle of incidence (which may be estimated from the angle of incidence of the waves generated at the paddles using Snell's law). A linear estimate of the incident and reflected time series can be obtained from the component amplitudes and phases of the incident and reflected wave spectra (e.g. Qijn, 1988). The techniques can be used over mild slopes and near to reflective structures as spatial variations in wave characteristics such as wavelength are small.

Mansard et al. (1985) use a least-squares fit of sine-waves to estimate reflection coefficients of first and second order components of regular waves. Maoxiang & Zhenquan (1988) extended Goda & Suzuki's method to three gauges to get set-down and long waves using low-pass filtering. However, only the use of filters to examine low frequency, long waves has become established in the analysis of laboratory experiments.

In wave flumes, it is preferred to suppress the impact of reflecting waves on the specified wave time series by active absorption. This requires wave measurements either very near to the wavemaker or wave gauges built into the wave paddle.

2.6.3   Wave skewness and asymmetry

In deep water the time series of surface elevations at a point is commonly assumed to be Gaussian, with a Rayleigh distribution of wave heights. As waves enter shallow water, their shapes evolve from sinusoidal to having short, high wave crests separated by broad, flat wave troughs. These waves, which are symmetrical about their vertical axis but non-symmetrical about the horizontal axis, are described as having skewness (of their surface elevation). As waves continue to shoal and break, they develop a pitched-forward shape with a steep front faces. These broken, sawtooth waves are described as having asymmetry, but near zero skewness (of their surface elevation) and are symmetrical about their horizontal axis, but not their vertical axis. Skewness and asymmetry are caused by nonlinear interactions between different frequency components of the wave spectrum during shoaling.

The kurtosis may also be used to identify non-Gaussian characteristics of a wave field. The kurtosis is defined by the fourth order moment of the surface elevation divided by the second order moment squared minus 3. For Gaussian sea states the kurtosis is zero.

The skewness of the water wave can be calculated as the third moment (i.e. mean of the cube) of the surface elevation normalised by the third power of the standard deviation (see the IAHR list of sea-state parameters for the formula). Elgar & Guza (1985) showed that the asymmetry could be calculated in a similar way to the skewness from the Hilbert transform of the surface elevation. Skewness and asymmetry can be calculated as the normalised third moments of the real and imaginary parts of the Hilbert transform (calculated in Matlab, for example).

The asymmetry is the same as the skewness of the slope (time-derivative) of the time series. Hence, for example, the surface elevation asymmetry is the same as the skewness of the surface slope (time derivative of surface elevation) which was called atiltedness by Goda (1986). More importantly, the asymmetry of the near-bed wave orbital velocity is the same as the skewness of the fluid acceleration (often referred to as acceleration skewness).

The skewness and asymmetry can also be investigated by calculating the bi-spectrum (Hasselmann et al., 1963; Elgar & Guza 1985; Elgar et al., 1995; Peng et al., 2009). The bi-spectrum gives a measure of the nonlinear interactions between three wave frequencies (with the third frequency commonly the sum of the first two) and is used to investigate self interactions (where the first and second frequency is the same) and sum and difference interactions. Wave skewness and asymmetry can be calculated from the real and imaginary parts of the double-integrated bispectrum normalised using the third power of the root-mean-square of the surface elevation (Hasselmann et al., 1963; Elgar & Guza 1985; Peng et al., 2009). Some research has also been conducted using the tri-spectrum (Elgar & Chandran, 1993).

2.6.4   Measurements of multi-directional waves

Along with the growing use of multi-directional waves (3D waves) in physical model tests, the requirement for measurements and analysis procedures increases. A well-established measure of 3D waves is their directional spectrum. However, most techniques for directional spectral analysis suffer from limited resolution. As an example, it is probable that incident bi-modal wave conditions (two primary wave directions in the same frequency range) as generated by a 3D wavemaker may in reality be closer to the target than estimated by the 3D spectral analysis. Increased resolution of the 3D spectral estimate can be obtained by measuring wave data within a larger area. This, however, has the disadvantage that the result represents average conditions over the footprint of the measurements. This is a problem for wave conditions with a large spatial change as would often be found in areas with significant refraction, diffraction and reflection as typical for coastal problems.

The general picture is that intrusive single point gauges for surface elevation and fluid velocity are routinely used at most labs. The present research interest is directed towards non-intrusive techniques. For point measurements, this is mainly motivated by the need for gauges mounted on moving carriages. Methods under consideration are radar level gauges, ultrasonic wave probes and laser beam with a video imaging technique. There is also a wide interest in measurements over a line or over an area, since arrays of point gauges are too expensive or too intrusive for this purpose. Some experience with Particle Image Velocimetry (PIV) is reported.

The problem of estimating directional wave spectra is usually simplified by the assumption that the 3D sea state may be represented by a 2D energy spectrum multiplied by a spreading function. The papers by Benoit (1993), Benoit & Teisson (1995) and Benoit et al. (1997) review the main methods for estimating directional spectra. Teisson & Benoit (1994) compare the performance of many of these methods in dealing with an extensive set of experiments on the reflection of irregular oblique waves off a rubble mound breakwater (with similar tests being used for rubble mound breakwater stability; Galland, 1994). The methods include:

- Direct Fourier Transforms, where the cross-spectra from two or more pairs of gauges are used. This method is reviewed by Goda (1985) and includes truncated and weighted Fourier decomposition as in Borgman (1969). A limited number of components can be resolved so directional spreading can only be crudely determined.
- Parametric models, which assume a specific formula for the directional spreading (e.g. cos2) which may be more useful in the lab where there is more control over the input spectrum than in the real sea.
- Variational Inverse Techniques (Long & Hasselmann, 1979).
- Eigenvector Method (Marsden & Juszko, 1987).
- Fourier Vector Method, which computes amplitudes, phases and directions from measurements of sea surface elevations and waves orbital velocities (Sand, 1979; Sand & Lundgren, 1979).
- Maximum Likelihood Method (MLM) which seeks to minimise the variance of the difference between the estimated and true spectra (Lundgren & Klinting, 1987; Krogstad, 1988, Isobe et al., 1984; Isobe, 1990). Yokoki et al. (1994) use

MLM with a reflected wave component in a similar way to Hughes (1993b). A known angular spreading function is used but this should not be a problem in the laboratory where the spreading function may be chosen. Isobe & Kondoh (1984) developed the Modified Maximum Likelihood Method (MMLM), which was the first method to evaluate the directional wave spectrum accurately in the presence of reflections, although this requires reflection line to be input. Davidson *et al.* (2000) developed an extension to the MMLM that calculated the position of the reflection line using a process of iteration. Both these methods break down and start to produce spurious peaks in the spectra when the measurement array is further away from the reflecting structure. The spurious peaks are caused by uncorrelated noise at frequency / direction pairs that have partial nodes at the location of the measurement gauges.

- Maximum Entropy Method (MEM), is an iterative method that maximises the entropy at each frequency. It is reviewed by Nwogu *et al.* (1987) and is supposed to give even better directional resolution than the MLM. Derived by Hashimoto & Kobune (1988) and Lygre & Krogstad (1986) using different definitions of entropy. Hashimoto *et al.* (1993) extended the MEM so that it can be used with arbitrary arrays of measuring instruments. It gives the same result as the MEM for 3-quantity measurements and similar results to the Bayesian method when used with more than three quantities and is a robust method as it allows for errors in the cross-power spectra.

- Bayesian Directional Method (BDM), as derived by Hashimoto & Kobune (1988). This takes into consideration errors in the cross-spectra and is computationally expensive, but is capable of evaluating the directional spectra of waves close to reflective structures by optimising a hyperparameter.

Benoit *et al.* (1997) concluded that the stochastic methods such as the MEM, MLM and BDM (and their variations) offer superior resolving power to the other methods. Ilic *et al.* (2000) and Chadwick *et al.* (2000) evaluated the MLM and BDM against numerical and field data including partial reflections and concluded that both methods could be used to determine incident and reflected wave fields when $T_w/T_s > 0.5$, where T_w = the time for the wave to travel to the reflector and back and T_s = length of time series used. Overall, the BDM was considered to be the more accurate method. Use of stochastic methods such as the MEM, MLM and (particularly) the BDM is recommended for the analysis of directional waves in a physical model, in situations that are more complicated than quasi-2D.

2.7 DATA MANAGEMENT

Over the last few decades there has been a simplification from unevenly distributed data sources to an orderly centralized data acquisition, although this is now becoming increasingly complex.

Before the 1980s – in absence of computers, data recording was done by observation, memos, paper plotter or tape recorder. Additionally the measurement installations were often widespread located along the test facility. Time synchronisation and most different calibration factors, e.g. at the same location separately for the gauge and the plotter, had to be considered. All this information had to be noted in a centralized test-journal. Often in the absence of any standards, the quality of this log hinged on the individual running the tests. The loss of the master engineer or the log was synonymic with the loss of the measured data for later interpretation.

Computer-based data logging offered for the first time a comfortable platform for synchronisation and standardisation of the data management. A powerful A/D-converter collects the data from all gauges in one point and at the same time. At this time nearly all devices were based on analog output signals and the measurements became congruent automatically. Incompatible sensors like videos, PIV or LDA were ignored and handled separately.

Today digitalization is not limited to the computer. Gauges are also digitised and deliver data in their own manufacturer-dependent data formats. Examples are the acoustic velocity meter (Vector/Vectrino), backscatter sensors or 3D-surface recordings. Again the data sources are decentralised located and must be synchronized in time and format. As an enhancement of complexity not even a calibration is available in time as just happened at the GWK for the sediment concentration sensors.

The aim of data management is to give a comprehensive frame of information together with the measured values both for automatic and public usage. This contains metadata such as a description of the tested structure, the flume layout and the seiching along and across the basin, the wave characteristics including the generation process and last but not least a complete description of each individual measuring device including localisation and calibration. Additionally the results of subsequent analyses should be integrated easily into the existing scheme at any time.

The required information accumulates at different time segments. The information is:

- The description of the tested structure more or less static for all test runs;
- The wave features which are changing for each individual test;
- The position and calibration for each measuring device, more or less static for all test runs;
- The statistics for each recording and device.

Due to the wide range of possible variations from test to test, a practical solution is to store all information for each single measurement, metadata and test condition, together with the actual measured data of any individual test. A simple solution is to establish one centralizing program, for example the data acquisition serving the A/D-converter, which asks for the wave parameters, starts the measurement, stores the raw-data and mixes the static information as a copy from a provided metadata-file with the statistics of the actual test run. A simple flow chart describes this in Figure 2.4.

Base of the information exchange files can be simple ASCII or Excel®- csv-formatted files. The information can be inserted with standard applications like an editor or Excel. These files can then be easily used by the individual lab-applications for data acquisition. A simple file structure with separate directories for each test-run was approved as most flexible. The raw-data files can be handled as separate files containing one-dimensional time series. 3D-gauges for example will produce 3 single files.

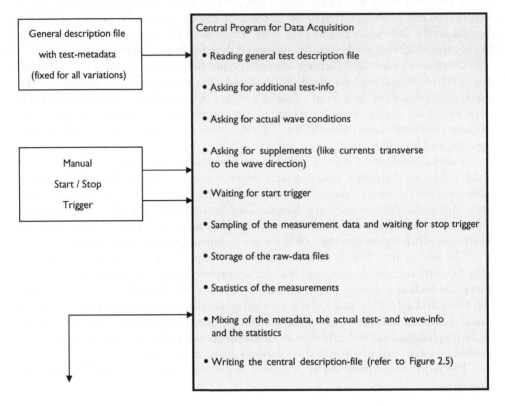

General description file
with test-metadata
(fixed for all variations)

Manual
Start / Stop
Trigger

Central Program for Data Acquisition

• Reading general test description file

• Asking for additional test-info

• Asking for actual wave conditions

• Asking for supplements (like currents transverse
  to the wave direction)

• Waiting for start trigger

• Sampling of the measurement data and waiting for stop trigger

• Storage of the raw-data files

• Statistics of the measurements

• Mixing of the metadata, the actual test- and wave-info
  and the statistics

• Writing the central description-file (refer to Figure 2.5)

Figure 2.4 Simple flow chart for preparation, measuring-process and storage of the test-log-file and raw-data (From FZK).

The presented technique can be extended at any time by enlarging the combined description file and adding the raw-data files into the directory.

2.8   TYPICAL SOURCES OF ERRORS

The following list covers the important sources of errors but is by no means exhaustive.

• Improperly placed boundaries;
• Artificial depth variations in front of wavemaker;
• Reflections;
• Limited test programme;
• Scaling of sediments;
• Permeability / porosity of model structures;
• Wave generation;

Worksheet for Physical Tests at GWK

global test description:	Cylinder groups in front of a caisson breakwater
user	Leichtweiss Institute Braunschweig, Prof. Oumeraci
flume information:	Large Wave Channel (GWK) of the FZK Hannover
flume size used l/w/h [m]:	227 5
channel geometry [m]:	0,0; 55,0; 62,1.2; 220,1.2; 222,2.2; 222,5.5; 227,5.5;227,2.2;
water depth [m]	4.75
wave height [m]	1.1
wave/peak period [s]	6.36
description of waves	
generation program	Jonswap-TMA-Spektrum
linear/nonlinear	SP-SET for pusher type wave generator
	linear
spectrum repetition time [s]	6000
no. of waves in rept. Time	986
absorption coefficients	0.92 0.85 250 205 250
start/stop ramping [s]	20

							Longitud. axis	Transverse ax	Altitude					statistic of the measurement			
channel no.	sample f [Hz]	start delay [s]	device	specific description	accuracy	dimension	X-position [m] 0 bis 330 (m)	Y-position [m] 0 bis 5 (m)	Z-position [m] 0 bis 7 (m)	linear-cal	x*1 cal-factor	x*2 cal-factor	x*3 cal-factor	min	max	DC	mean
1	120	0	wave gauge	single wire	0.01 m	m	0.04	2.5	0	0.0231				-0.7	1.12	-0.3	0.21
2	120	0	wave gauge	single wire	0.01 m	m	3.59	2.5	0	0.0231				-0.67	1.16	-0.25	0.14
3	120	0	wave gauge	single wire	0.01 m	m	49.5	0.5	0	0.0231				-0.81	1.25	0.02	0.11
4	120	0	wave gauge	single wire	0.01 m	m	53.1	0.5	0	0.0231				-0.84	1.24	-0.04	0.08
5	120	0	wave gauge	single wire	0.01 m	m	58.22	0.5	0	0.0231				-0.8	1.26	-0.11	0.12
6	120	0	wave gauge	single wire	0.01 m	m	66.25	0.5	0	0.0231				-0.79	1.28	0.11	0.07
:																	
17	120	0	wave gauge	single wire	0.01 m	m	210	0.5	0	0.0231				-1.15	1.25	0.02	0.11
18	120	0	wave gauge	single wire	0.01 m	m	221.85	0.5	0	0.0231				-1.18	1.24	-0.04	0.08
19	120	0	strain gauge	x-dir. bending	0.5 kNm	kNm	102	2.35	6.5	30.543				-5.45	10.33	3.5	0.14
20	120	0	strain gauge	x-dir. bending	0.5 kNm	kNm	102	2.35	6.5	30.543				-5.45	10.33	2.84	0.11
21	120	0	strain gauge	x-dir. bending	0.5 kNm	kNm	102	2.65	6.5	30.543				-5.45	10.33	3.02	0.08
22	120	0	strain gauge	y-dir. bending	0.5 kNm	kNm	102	2.65	6.5	30.543				-5.45	10.33	2.98	0.12
23	11000	100	pressure cell	Druck PDCR 8:0.6 kPa		kPa	222	2.25	2.5	0.00103				0	15.3	0	1.45
24	11000	100	pressure cell	Druck PDCR 8:0.6 kPa		kPa	222	2.5	2.5	0.00103				0	16.78	0	1.05
25	11000	100	pressure cell	Druck PDCR 8:0.6 kPa		kPa	222	2.75	2.5	0.00103				0	18.14	0	0.72
:																	

The information of the medium grey highlighted arrays are imported from the general description file containing the test metadata

The information of the light grey highlighted arrays are given by the technician just before starting the measurement

The information of the dark grey highlighted arrays will be calculated automatically by the data acquisition program in the end of the measurement

Figure 2.5 Example of an Excel®-csv-formatted info-log file containing all relevant information of a single measurement, can be extended (From FZK).

- Duration of test;
- Selection of sea states for modelling;
- Method of analysis;
- Measurement of reflected waves;
- Resonance.

2.9   FUTURE WORK

Topics for future studies could include the following:

- Generation of irregular waves with and without non-linear control;
- Generation of extreme (freak) waves;
- Reflections from large structure in basins and flumes;
- Floating structures in irregular waves on shallow water (addressing free and bound long waves).

Chapter 3

Breakwaters

Guido Wolters

William Allsop, Luc Hamm, Doris Mühlestein, Marcel van Gent, Laurent Bonthoux, Jens Kirkegaard, Rui Capitão, Juana Fortes, Xavi Gironella, Liliana Pinheiro, João Santos & Isaac Sousa

3.1 INTRODUCTION

In coastal design analytical, empirical and numerical methods are used to compute the hydraulic and structural performance of a coastal structure. These involve a certain degree of simplification of the real situation. Furthermore, there are no typical construction patterns that are used; the design of structures is ever changing. Each construction is unique in the fact that it is designed according to the local boundary conditions (e.g. bathymetry, wave climate, available material). Therefore, it is also very difficult to transfer design-/model experiences from one construction to another. If earlier design-/model experience is available for the structure and if the constructions do not deviate greatly in structural and hydraulic boundary conditions, analytical and numerical methods are usually regarded as sufficient in the design. However, the further the structure or design conditions depart from previous experience or from the idealised configurations on which the empirical design methods are based, the larger the uncertainties associated with using these methods. For large and complex hydraulic structures, physical models therefore become necessary in order to achieve a final design that is sufficiently accurate and economically optimised.

Physical model tests are required where the importance of the assets being defended (and/or of the structure itself) is high or when the stability of the structure is not guaranteed by analytical models/semi-empirical formulas and numerical models. They are also employed when:

- Designs have to be optimised;
- Overtopping is a major parameter of the study;
- Complex phenomena such as wave breaking and wave transmission are analyzed;
- The bathymetry or the structure geometry is complex;
- Transitions between structures / structure sections are to be studied;
- Concrete armour units are employed as primary armour (especially those with potentially more brittle failure mechanisms and if a reliable quantification of small armour movements is important).

This chapter deals with physical modelling of short term, short wave models of fixed rubble-mound breakwater structures under wave influence. The following items are not discussed in this chapter:

- Other breakwater structure types than rubble mound structures (e.g. impermeable structures, vertical walls, composite structure etc.), however, much of what is written in this report is also valid for other structural types
- Movable-bed physical models (and scour): movable-bed models have a bed composed of material that can react to the applied hydrodynamic forces. The scaling effects inherent in movable-bed physical models used for studying sedimentary problems are not as well understood as they are for fixed-bed models. Consequently, movable-bed model results must be carefully reviewed in the context of previous, similar models that have demonstrated success in reproducing prototype bed evolutions.
- Structure-foundation interactions
- Long-wave models: these are based on similar modelling procedures, although the relative importance of individual scaling effects is different (e.g. the reflection from boundaries becomes more important due to the higher reflection coefficient of long waves; active absorption systems have more difficulty with longer waves, alongside other problems due to small scales and low wave heights)
- Long-term effects: these describe system changes that occur over extended time periods (days to years) and are not as typical as short-term models. Short-term models are typically used to describe the model response to a design condition with a particular storm duration.
- Effects of currents
- Numerical models and their limitations

This document forms an introduction into the field of physical modelling of breakwater structures. The following references and guidelines, which are widely accepted in the field of physical breakwater modelling, are used as underlying basis for this document:

- Hughes S.A., 1993
- Dalrymple R.A., 1985
- TAW, 2002
- CIRIA, CUR, CETMEF, 2007
- British Standards Institution, 1991
- Owen M.W. & Allsop N.W.H., 1983
- Owen M.W. & Briggs M.G., 1985

3.1.1 Purpose of physical model testing

Scale models of rubble-mound structures are typically employed in order to (e.g. Hughes, 1993a):

- Investigate structure characteristics such as wave run-up, rundown, overtopping, reflection, transmission, absorption, and static/dynamic internal pressures for different structure types, geometries, and/or construction methods;

- Examine the stability of rubble-mound armour layers protecting slopes, toes and crowns when exposed to wave attack;
- Determine the hydrodynamic forces exerted on crest elements by wave action;
- Optimize the structure type, size and geometry to meet performance requirements and budget constraints;
- Examine alternate construction sequences under different wave conditions;
- Determine the effects that a proposed modification might have on an existing structure's stability and performance;
- Develop and/or test methods of repairing damage on existing structures or improving the performance of an existing structure.

Despite the shortcomings inherent in experimental physical modelling, scale models are the best tool that an engineer has, to discover and verify engineering solutions for breakwaters and to improve upon and verify mathematical and numerical approaches.

3.2 MODEL SET-UP AND OPERATION

3.2.1 Choice of physical model

A physical model is a 'reproduced physical system (usually at a reduced size) so that the major dominant forces acting on the system are represented in the model in correct proportion to the actual physical system' (Hughes, 1993a). The choice of the physical model depends on various project- and site specific parameters:

- Model test objectives;
- Proposed structure design (and rationale behind the proposed structure). These include typical cross sections, detailed drawings of the structure and material related information such as density of materials and grain size distributions;
- Bathymetric details of the surrounding area (3D) or the wave approach direction (2D) so that a representative bottom configuration can be constructed in the model. Good quality bathymetric data will assure that wave transformation is well simulated in the model;
- Environmental design conditions such as wave height, wave period, water levels, wave spectra and wave directions;
- Structure performance criteria such as allowable damage level, maximum wave run-up, permissible wave transmission at design wave and water level conditions;
- Any other criteria pertaining to the structure's purpose and function such as e.g. structure construction sequence, tolerances and placement specifications.

These parameters must all be considered to determine the required model size (flume/basin size and necessary water depth), whether a 2D or 3D model set-up is required, the scale of the model and the required wave paddle capabilities (frequencies, reflection compensation). For typical types of model facilities, refer to HYDRALAB (2007a).

It is the responsibility of the client to use appropriate safety factors for the modelling results according to national guidelines.

3.2.2   Layout of the model

Independent of the chosen model scale is the physical model affected by the artificially introduced partitioning of the prototype situation in the model. Often, only a section of the prototype situation is modelled. In general the model positioning in the wave channel/basin should be such that boundary effects are minimized and that the given wave conditions are achieved over the appropriate test section(s) and produce the appropriately scaled responses. 3D models are especially susceptible to reflections that propagate from the model boundaries into the measuring area, falsifying the measurements. By using wave dampers, these reflections can be reduced. To allow the modelling of varying wave directions, guide walls can be used in order to prevent artificial diffraction effects (Figure 3.1), such as the spilling of energy into areas where the wave machine cannot accurately reproduce the wave climate. The guide walls should be placed parallel to the wave direction, so that wave reflections from the guide walls do not distort the wave field.

A correctly reproduced bathymetry ensures that realistic wave conditions such as wave dispersion, wave refraction/diffraction, shoaling and wave breaking are correctly simulated. The bathymetry is usually not modelled in every detail; rather the main bottom contours (isobaths) are represented. Special attention should be paid to the bottom contours within 1–2 wavelengths (1 L–2 L) from the structure toe, since those have a significant influence on the wave climate and loading conditions at the structure (Van Gent & Giarrusso, 2005). A correct representation of the bathymetry

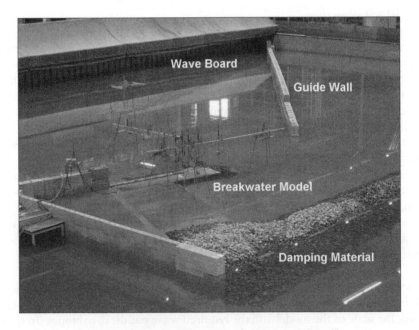

Figure 3.1 Typical breakwater layout (using a guide wall).

is especially important where features such as a dredged channel or rock outcrops may focus wave action (3D).

Seabed roughness is usually not taken into account. It can however become important at shallow water depths. Depending on the dimensions of the wave basin, a transition slope is often used between the modelled bathymetry and the deeper water section at the wave paddle to allow proper wave generation. The transition slope should have an inclination of ≤1:10.

The length of the 'deeper water' section between wave paddle and transition slope should be larger than 3–5 times the water depth h at the wave paddle (based on Dean & Dalrymple, 1991; Biesel & Suquet, 1951). This ensures that the evanescent wave modes near the wave paddle have decayed. The distance is usually enough to ensure the necessary space for the wave probe array to assess incident and reflected waves near the wave paddle. Usually a length of 3–5 m fulfils these conditions, depending on the chosen scale. The 'deeper water' section should guarantee a minimum water depth of $h/H_s > 3$ (e.g. CIRIA, CUR, CETMEF, 2007).

For horizontal bathymetries multi-gauge arrays should preferably be located at a distance of more than 0.4 times the peak wave length L_p from the structure to improve incident wave assessment (less effects from evanescent modes (Klopman & Van der Meer, 1998). Goda (2000) specifies a distance of 1.5 times the local wave length. For the length of the modelled bathymetry section (after the transition slope), usually a distance of >3–5 times the local wave length L is recommended to allow the accurate modelling of shoaling and breaking of incident waves according to the local bathymetry.

Common methods for constructing fixed-bed bathymetry consist of making bottom elevation templates or level contours (from wood/masonry etc.) that are affixed to the floor of the wave facility. The space between the templates is filled with sand (or gravel) and capped with a concrete layer. Fixed-bed bathymetry will usually not adversely affect armour layer stability studies because most damage occurs in the vicinity of the still water level. Thus the problem of correctly scaling the foundation/bed material can be avoided. However, studies of toe protection stability should consider the impact of modelling on a fixed-bed. Moving-bed bathymetry could be considered if scour is judged to be a major problem (particularly at the breakwater toe). Furthermore, if active morphological features are present, like mud banks or moving bars, their influence on wave action should be taken into account. However, their influence on structure loading is likely to be lower than the effect of the bottom configuration (low-tide terrace) closer to the structure (Van Gent & Giarrusso, 2005).

Bathymetry changes can be expected during storms. Thus an alternative to the use of measured profiles in physical modelling could be the use of an adapted

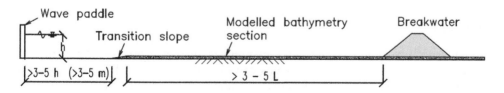

Figure 3.2 Bathymetry layout.

profile, based on scour and sediment transport estimations, which could present the bathymetry during a storm more accurately. Alternatively, as a conservative estimate, a larger water depth could be used. There are very few cases where the bathymetry can safely not be represented, for example in the case of small waves and at larger water depths. Modelling of the bathymetry can be neglected if all of the following conditions are fulfilled:

- $\tan\alpha < 1{:}250 - 1{:}1000$
- $h/L_0 > 0.045$
- $H_s/h < 0.3$

where $\tan\alpha$ describes the foreshore slope, h/L_0 is the relative water depth on the foreshore (L_0 = deep water wave length) and H_s/h is the relative wave height on the foreshore. These conditions correspond to an Ursell number of $U < 150$. If one of the conditions is not fulfilled it is recommended to model the bathymetry.

3.2.3   Structure

Solid structures are usually constructed in metal, timber or concrete whereas rubble mounds are made of sieved or weighed quarry rock. During the model construction (wooden) profiles of the structure cross-sections are used with mutual distances depending on the structure. Structural elements (e.g. when loading of crest elements is investigated) should be rigid enough to prevent or minimize unwanted structural oscillations. If unwanted structural oscillation cannot be prevented it should be in a frequency band outside the typical wave spectrum (usually $f < 0.01$ Hz or $f > 10$ Hz).

Crown walls

The modelling of breakwater crown wall stability (sliding or overturning failure) can be done by various methods:

1   Whole body forces can be measured on a section of crown wall using a force table or array of force elements.
2   Wave pressures on the front face and underside can be summed to determine the whole body forces and moments.
3   The crown wall section can be reproduced at a reduced weight so that the friction forces between armour and crown wall are in similitude with the prototype forces.

Method I

This approach is often used for the measurement of global horizontal forces. The quality of the uplift force measurements can however be diminished: The use of a force table or similar device requires that the model crown wall element is suspended so as not to touch the underlying rock. The introduction of any such gap may alter wave up-lift pressures acting on the underside of the crown wall, both in magnitude and distribution. Use of permeable foam to reduce flows in this gap might be sufficient if its flow/pressure transmission characteristics are appropriate, but this has particular uncertainties.

Method 2

This approach has the advantage that both pulsating and impulsive loads will be recorded at the location of the pressure transducers (if the sampling rate is high enough). A disadvantage of this method is that pressures are only measured locally and then multiplied over the whole representative area to provide a force estimate. In contrast to Method 1 the reliability of the uplift forces is higher.

Method 3

This approach is very simple and may therefore be useful for rapid assessment of safety against sliding and overturning, but does not identify the loads *per se*, only the onset of sliding.

3.2.4 Waves

Due to the nature of physical model testing, wave parameters and wave related information are critical to a successful modelling approach. For hydraulic model testing in waves refer to chapter 2 for a comprehensive overview of aspects concerning waves that are not in the remit of breakwater modelling in particular:

- Wave generator types and wave generation (actuators, control signals);
- Long waves;
- 2D/3D effects regarding waves;
- Multidirectional waves and directional spectral analysis;
- Wave reflection & wave trapping.

Wave parameters

In the fixed design model, water levels are usually coupled with wave conditions. High tide levels are often used to test the upper part of the structure, whereas the toe stability is often tested against low tidal levels. It should however, be kept in mind that extreme water levels are not always the most critical design condition in stability testing.

Design wave conditions are usually provided for different return periods (typically between 1 and 100 years return periods) including the significant wave height, the peak or mean wave period, the peak or mean wave direction and the duration of the storm (or number of waves). In order to produce a statistically reliable test result, duration have to exceed a certain minimum number of waves. This is usually between 500–3000 waves. Typically, applied storm durations of 3–6 hours (in prototype) satisfy this condition. Storms are often simulated as a series of test runs with fixed wave conditions, increasing in severity depending on the likely storm profile.

Wave energy spectra in physical models are usually characterized by their spectral parameters. The most commonly used are the significant wave height H_{m0} and the peak period T_p. Other spectral parameters are also used. For example, to give a better approximation of nearshore conditions, the spectral wave period $T_{m-1,0}$ ($T_{m-1,0} = m_{-1}/m_0$), T_{01} ($T_{01} = m_0/m_1$), T_{02} ($T_{02} = (m_0/m_2)^{1/2}$) are used. Time domain parameters such as the average zero crossing wave period T_z, the wave period of the largest 1/3 of waves $T_{1/3}$, the significant wave height H_s (or $H_{1/3}$) and the wave height of the largest 1/10 of waves $H_{1/10}$ are also common.

The most commonly employed wave spectra are JONSWAP (confined young seas) and Pierson-Moskowitz (PM, fully-developed open seas). Depending on the local conditions also other spectra are used (e.g. TMA in shallow water, Bretschneider). For JONSWAP spectra usually a peak enhancement factor of $\gamma = 3.3$ is applied ($\sigma_a = 0.07$, $\sigma_b = 0.09$). The frequency spectrum should be defined together with a directional spreading function in 3D wave tanks. For further information refer to the sections on model operation and storm profile modelling in Chapter 2.

Wave generation

Most commonly employed wave boards can generate regular/monochromatic and irregular/random waves. Wave energy spectra can be prescribed by using standard or non-standard spectral shapes or by a specific time-series of wave trains. The wave board is often equipped with an active reflection compensation system. This means that the motion of the wave board compensates for the reflected waves, preventing them to re-reflect towards the breakwater model. This is usually the case in 2D wave channels. In 3D facilities only multi-paddled wavemakers can be equipped with this system. If no reflection compensation is available, and presuming that the set-up allows for non-parallel placement of breakwater and wave-board, the breakwater structure orientation should be such that reflection occurs towards the wave dampers and not back towards the wave board. Reflection compensation becomes especially important when reflection coefficients are large, such as in the case of vertical walls, steep beaches / cliffs and where there is little room for passive absorption. It is however, limited by the capacity of the wave maker. Reflection compensation is always important in the case of long waves. These have almost full reflection from spending beaches or structures and may resonate without reflection compensation.

Various commercial and in-house software is used for the wave generation (for instance Delft AUKE /generate (Delft Hydraulics), Wave Synthesizer (DHI), GEDAP/AWA, WAVELAB2 ((Aalborg University), SAM) and post-processing of the model results. Generally specifically written Matlab and Fortran routines are used to supplement this software. First order wave generation (based on linear wave theory) is typically used. Second order wave generation, based on the 2nd order theory from Stokes (1847) and Longuet-Higgins & Stewart (1964), as incorporated in the AUKE, DHI and GEDAP software (Mansard, 1991), is an option if a more accurate representation of waves in nature is needed. Second order wave generation means that second order effects of the first higher and first lower harmonics of the wave field are taken into account in the wave board motion (Klopman & Van Leeuwen, 1990; Van Leeuwen & Klopman 1996; Van Dongeren et al., 2001; Schäffer & Steenberg, 2003).

Second order wave generation is especially valuable if deep water conditions cannot be achieved at the wave paddle. In this case, 2nd order wave generation guarantees a better representation of the wave shape and less spurious waves. The first order transfer function between water elevation and wave paddle motion is described in Dean & Dalrymple (1991) and Biésel (1951).

Wave absorption

Besides the (active) wave damping already performed at the wave paddle (see section 3.5.3) (passive) wave absorbers are also placed at the end of a 2D wave channel or around a 3D model. Generally, gentle slopes of crushed stone/shingle are used in combination with parabolic shaped slabs (perforated and/or with ripples) and perforated screens/permeable mats. A good wave damper should show the same damping behaviour, independent of wave steepness. For a survey of available techniques refer to Ouellet & Datta (1986, also for parabolic beaches), Jamieson *et al.* (1989) Twu & Liu (1992), Twu & Lin (1991) for perforated screens.

Wave calibration

A careful wave calibration exercise is recommended wherever a complicated foreshore or structure set-up is used, or where wave conditions at or near the intended structure are to be achieved. Where the foreshore is simple/shallow, and/or wave conditions are imposed near the wave paddle, simplified wave calibration may be sufficient. It is convenient to differentiate between modelling in 2D (wave channel) and 3D (wave basin). In 2D models, boundary effects should be relatively small so that a reliable incident wave spectrum may be obtained without a detailed wave calibration exercise if an active reflection absorption system is used. This method requires that waves at the measurement devices are not corrupted by intermediate or shallow water wave breaking.

A careful wave calibration exercise is recommended where a complicated foreshore or structure set-up is used in a 3D model and wave conditions at or near the intended structure are to be imposed. This exercise requires measurements of wave conditions in the absence of the structure so that the design incident wave conditions are achieved at the toe of the structure. The absence of the reflective structure will permit better estimates of incident waves at the position of the (planned) structure. Model observations with the structure in place can then be related to the incident wave conditions without structure. The calibration step may not be essential if measurement equipment is used that can separate incident and reflected waves, in combination with a proper prediction of the wave field that needs to be generated to achieve the required conditions at the structure. If no calibration step is used, an iterative procedure is often necessary to achieve the desired target wave spectrum at the structure toe.

Wave type

For design purposes the loading of coastal structures should be modelled using irregular (also known as random) waves.

Both long and short crested waves are employed in physical modelling. Long crested waves are generally believed to give conservative results (larger energy input) for damage and wave overtopping. Short-crested waves are important if local wave characteristics/phenomena around the breakwater are to be investigated (Herbers & Burton, 1997), and may give more severe local effects (Abernethy & Ollover, 2002). For wave penetration into harbour basins, short-crested waves may be preferred where long-crested conditions may give less complete results.

Wave separation and transmission

Incident waves are usually assessed by separating (spectrally) the incoming waves into incident and reflected waves. Typically employed techniques for a near-horizontal bottom and uni-directional waves are e.g. Mansard & Funke (1980), Goda & Suzuki (1976), Frigaard & Brorsen (1995). They are based on measuring the incoming waves at several closely spaced locations. Three-point or more point techniques (5–9 points) are typically used. Two-point techniques are less reliable, since they resolve only wavelengths that are not a multiple of twice the distance between the two wave gauges. Moreover, these methods also exhibit a lower investigable frequency range, higher sensitivity to noise and deviations from linear theory. If a two-point technique is used, care must be taken in the analysis of the results. Three-point techniques are typical for 2D models, whereas 3D models use more points. Another option in 3D is the use of directional wave gauges (e.g. wave gauge/pressure sensor coupled with currentmeter) which resolve the incident wave spectrum based on the particle velocities in two directions and the change in surface elevation (Maximum entropy method, e.g. Massel, 1998; Massel & Brinkman, 1998). For a comparison of sensor array and co-located gauge (gauges located on the same vertical line) performance in a flume refer to Hughes (1993b). Transmitted wave heights can be determined in the same fashion, although usually only standard wave gauges are employed. It is stressed that to guarantee accurate results, proper wave absorption must be implemented.

The described techniques are applicable for horizontal and gently sloping foreshores, where the wavelength does not vary significantly over the length of the wave gauge array. For more inclined (non-horizontal) foreshores or long waves, wave separation should be based on Baldock & Simmonds (1999, for steeper foreshores) and Battjes et al. (2004, for long waves).

Multiple wave directions

The rubble mound structure is usually loaded with waves from various directions. The varying wave directions can be realized by moving the wave generator, pivoting the model when the generator is fixed or optimising the breakwater location, such that the most important wave directions can be tested in the same model set-up (e.g. with a 'snake-type' wavemaker and independently controlled wave paddles) using guide walls if necessary. Simultaneous testing with multiple wave directions is done only rarely. Multi-directional facilities of this kind are an exception. Various wave directions are usually tested in appropriate sequences instead, using a 3D set-up. This is suitable for most situations. For more information on multidirectional waves and directional spectral analysis, refer to the wave modelling guidelines (HYDRALAB, 2007a).

Waves and currents

Current influence under waves is not usually considered in breakwater stability modelling. However, physical modelling should be considered if the current is strong enough to significantly affect the toe stability of the structure. Currents could also play a key role in breakwater stability when they flow opposing to the waves with a current velocity to wave velocity ratio (u/c) higher than 10%. This happens at tidal inlets or is caused by rip-currents which are guided by groynes or similar structures. Such a case is described for example in Usseglio-Polatera *et al.* (1988), where breakwater damages were attributed to the increase of incident wave heights by a rip-current. This rip-current was estimated by numerical simulation leading to a peak velocity of 3 m/s with offshore wave conditions of $H_s = 7.1$ m and $T_p = 12$ sec.

2D/3D Models

Usually 2D models are used to optimize breakwater cross section(s) and 3D models to verify/optimise the roundhead and other 3D features of the breakwater. For the analysis of the interaction between waves and the trunk of structures often a 2D model is sufficient. 2D and 3D models are often also combined so that the weak interactions (refraction, diffraction) are modelled with the 3D model and the strong interactions between waves and structure with a 2D model at a larger scale. The 2D model is usually a typical cross section of the investigated structure at the point of maximum wave exposure respectively for the most important wave direction.

3D physical models are required whenever the structure is three-dimensional or the wave action at the structure is significantly oblique (say $\beta > 30°$, where β is the angle from the perpendicular), short-crested or focussed and in situations with very irregular seaward bathymetry. They are particularly important near to a dredged channel, at roundheads, elbows, connections to the coast, transitions between various types of armour, and other singular points or where the structure plan shape is complex. A 2D simplification may be safe to examine the performance of long trunk sections where obliquity is small ($\beta \leq 15°$) or where the design will be adjusted by using empirical design methods, calibrated by the model tests. 2D models can be used to validate typical sections quickly and conservatively. For further 2D/3D effects refer to the wave modelling guidelines (HYDRALAB, 2007a).

3.2.5 Scaling

Choice of scale

There are a number of factors that have to be taken into account for scale selection of a short-wave hydrodynamic physical model, and these factors vary widely from one modelling application to the next. In the broadest sense, the engineer must consider:

- Past experience with physical models of similar nature;
- Design wave conditions and water levels as compared to wave maker capabilities and available water depth;
- Seabed bathymetry;

- Area of interest as compared to facility/tank dimensions;
- Armour unit size and availability of model concrete armour units;
- Range of parameters to be considered in the model;
- Resolution of responses/measurements required;
- Flow Reynolds numbers for the proposed scale;
- Facility availability;
- Contract requirements;
- Funding limitations.

Usually a range of scales is chosen which comply with Froude similitude and then this range is narrowed down in accordance with the above requirements. Generally small scale models are used. However, it can become necessary to do large scale models (>1:10) or field tests if a correct representation of the conditions in nature cannot be guaranteed in a small scale model. A large model scale (large waves) is often also necessary for the accurate modelling of wave loading and structure responses. Large scale models are also used in research applications because of smaller/negligible scale effects. It is advisable to operate at larger scales whenever possible. However, the benefits have to be weighed against cost and time. Large scale models can be complicated, costly and are not always necessary.

Scaling laws

The necessary scale of a model is chosen in such a manner that all important wave conditions and structural parameters are reproduced to an adequate degree and a sufficient measuring accuracy is guaranteed. The scale of the model is determined by geometric, dynamic and kinematic similarity. Geometric similarity of a model is given when all geometric lengths L_p in prototype have a constant relation to the corresponding lengths in the model L_m ($n_L = L_p/L_m$). Kinematic similarity says that time-dependent processes in the model have a constant time relation to the processes in nature ($n_t = t_p/t_m$). Dynamic similarity entails that the forces in nature and model have a constant relation ($n_F = F_p/F_m$) with the premise that in geometrically similar models time dependent processes have kinematic similarity. Thus for a geometrically similar model, the key to a correct representation is dynamic similarity. No geometrical model distortions should be introduced where horizontal and vertical scales differ. This leads to the typically used scaling laws. Dependent on the importance of the individual attacking forces, various scaling numbers have been introduced such as Froude (*Fr*), Reynolds (*Re*), Weber (*We*), Mach (*Ma*), Cauchy (*Ca*), Richardson (*Ri*), Euler (*Eu*) and Strouhal number (*St*).
The most important are:

$$Fr = \frac{u}{\sqrt{g \cdot L}} \tag{3.1}$$

$$Re = \frac{u \cdot L}{v_k} \quad \text{or} \quad Re_D = \frac{\sqrt{gH_s \cdot D_n}}{v_k} \quad \text{(for flow conditions in the armour layer)} \tag{3.2}$$

$$We = \frac{\rho_w \cdot u^2 \cdot L}{\sigma} \tag{3.3}$$

$$Ca = \frac{\rho_w \cdot u^2}{E} \tag{3.4}$$

in which:
u = particle velocity (m/s)
g = gravitational acceleration (m/s²)
v_k = kinematic viscosity of water (= 10^{-6} m²/s) (m²/s)
ρ_w = density of water (kg/m³)
σ = surface tension (N/m)
L = characteristic length (m)
H_s = significant wave height (m)
D_n = nominal diameter of the armour units (m)
E = modulus of elasticity (N/m²)

For wave models, the relevant forces are the forces of gravity, friction and surface tension. For true dynamic similarity, the Fr, Re and We numbers must then be the same in model and prototype. This is, however, not possible. For a Froude – Reynolds model the ratio of the kinematic viscosity for the model and the prototype are fixed by the geometric scales. For most scales, it is impossible to find an appropriate model fluid and therefore true similarity is not possible. The importance of friction is often small however, since waves must propagate long distances before bottom friction seriously affects them and in the case of drag forces, there are ranges of Reynolds numbers where the drag coefficient is constant. If the modeller uses a Froude model law (since the wave field is dominated by the influences of gravity and inertia) and insures that the model Reynolds number is in the same range as the prototype, then the Reynolds number need not be exactly the same. This type of argument holds for the Weber number as well. Surface tension is generally negligible in prototype waves and therefore if the model is not too small (wavelengths must be much greater than 2 cm, wave periods >0.35 s, water depths >2 cm, Le Méhauté, 1976), Weber similitude can be neglected. Otherwise the model will experience wave motion damping that does not occur in prototype. A closer evaluation of surface tension effects can be done e.g. by the method described in Hughes (1993a). Thus, in coastal engineering usually the Froude law has to be fulfilled; surface tension, roughness and viscosity influences have to be considered in the above described manner. If these criteria are met, the influence of the Re and We numbers are usually negligible.

In the case of modelling flow through the structure and (drag) forces on structures, Reynolds similitude becomes more important. In this case it is very important to come as close as possible to the prototype Reynolds number and to model waves (and currents) accurately. However, if the Reynolds number is large enough and fully turbulent flow conditions in the armour-/filter layers are preserved (Re_D > 30000, e.g. Dai & Kamel, 1969) the forces can still be scaled by Froude alone. Furthermore, a minimum structure/stone size needs to be kept. Viscous effects can be limited in the model if diameters larger than 3–5 mm (in model scale) are employed. Also, in very shallow water, when wave breaking occurs, the bottom friction becomes relevant and needs to be adequately modelled (if necessary using artificially introduced bottom friction). For breaking waves also the Weber and Cauchy numbers can become important since they determine the air intake, water turbulence and water compressibility. Hence the model scale should not be chosen too small. Furthermore, compressibility is considered for wave loads on solid structures.

If sediment transport is modelled also other similarity criteria must be met (see chapter 4 on modelling of sediment dynamics). The same is valid for the assessment of current forces and current induced structural oscillations. Sediment transport and current induced effects are however not included in this guideline.

Similitude in rubble mound breakwaters

Similitude in rubble mound breakwaters depends on a variety of structural and physical parameters. In general, the following scaling criteria have to be fulfilled:

* Overall structural dimensions are scaled geometrically;
* Flow hydrodynamics (waves) need to conform to the Froude criterion;
* Turbulent flow conditions have to exist throughout the primary armour layer (satisfied reasonably by the $Re_D > 30000$);
* It is best to operate at larger scales when possible (since viscous forces can be greater if flow velocities and units are small).

From the Froude law, the following typically used scaling relationships, expressed in terms of the length scale factor n_L, can be derived:

Wave height (m)	$n_H = n_L$
Time (s)	$n_T = n_L^{0.5}$
Velocity (m/s)	$n_u = n_L^{0.5}$
Acceleration (m/s²)	$n_a = 1$
Mass (kg)	$n_M = n_\rho * n_L^3$
Pressure (kN/m²)	$n_P = n_\rho * n_L$
Force (kN)	$n_F = n_\rho * n_L^3$
Discharge (l/s/m)	$n_q = n_L^{1.5}$

Permeability scaling

In the underlayers and core of model breakwaters, geometric scaling of the material sizes may lead to viscous scale effects because these layers can become less permeable, thus limiting wave-driven flows into the inner layers and increasing the flow effects in the armour (Oumeraci, 1984). Geometric scaling of underlayers and core material could thus be regarded as a conservative estimate of armour stability (larger damage and overtopping in the model).

However, geometric scaling of the material sizes will lead to different values of transmission and reflection from what occurs at prototype scale (more energy reflected and less transmitted). Various methods are therefore available to scale for permeability of core and underlayer materials (Hudson *et al.*, 1979; Jensen & Klinting, 1983; Van Gent, 1995). The scaling of permeability is thereby usually achieved by equating the hydraulic gradient at the interface between rock layers in prototype and model (e.g. Jensen & Klinting, 1983; Van Gent, 1995). This can be done by equating the Forchheimer resistance of the materials. However, where the size grading of the underlayer and core materials are not well-established and cannot be assured in the construction, a distorted material size might give over-optimistic results for the stability analysis, so that it may be unsafe to apply such a permeability correction. Furthermore, sand incursion into the core can occur at prototype scale and result in drastically reduced permeability. In these cases geometrically scaled material can give unconservative results. For a discussion of scale effects regarding permeable structure parts refer to Cohen de Lara (1955) and Le Méhauté (1957, 1958). It can be seen that the benefits and disadvantages in permeability scaling must be carefully weighed, depending on the goal of the study and the client's needs and specifications.

Relative densities

When strict geometrical scaling is applied to the armour, the ratio of fluid mass density to the immersed mass density of the armour unit mass:

$$\Delta = \rho_a/\rho_w - 1 \tag{3.5}$$

should be the same both in model and prototype. A method for compensating for the increased buoyancy of the salt water relative to the fresh water used in most scale models is to adjust the weight of the model armour units. The scaling requirement is based on preserving the value of a 'stability parameter' between prototype and model (see Hughes, 1993a).

Stability scaling

Various methods exist for the stability scaling of materials such as Hudson (e.g. CERC, 1984) and Ahrens (1989, which includes the wave period). In stability scaling, it is ensured that the stability number N_s (Hudson) is the same in model and in nature. Stability scaling is of relevance for the toe material and the armour layers. The differences in water density (salt water in nature and fresh water in the model) and in the armour unit density are accounted for in this parameter. The stability number is defined as:

$$N_s = \frac{H_s}{\Delta D_{n,50}} \tag{3.6}$$

Where:

Δ	relative mass density $= (\rho_a - \rho_w)/\rho_w$ (–)
ρ_a	density of armour units (kg/m^3)
ρ_w	density of water (kg/m^3)
$D_{n,50}$	nominal diameter of the armour units, based on M_{50} (m)
H_s	significant wave height (m)

Thus the stability of the armour units is modelled correctly when the stability number in the model is the same as the stability number in prototype. This is the case when:

$$n_D = n_L / n_\Delta = n_L \cdot \frac{\Delta_{model}}{\Delta_{prototype}} \quad \text{(scaling relationship for armour diameter D)} \quad (3.7)$$

or

$$n_W = n_p \cdot n_D^3 = n_p \cdot \left(\frac{n_L}{n_\Delta}\right)^3 \quad \text{(scaling relationship for armour weight W)} \quad (3.8)$$

It is noted that absolute geometric similarity of the dimensions (diameter) of the model armour units is not necessarily maintained. This results in slight differences in the geometry of the armour layer, as compared to nature. Small differences do not significantly affect the results of the model experiments (effects on stability and overtopping are generally less than 5–10%) if care is taken that the armour geometry is correctly reproduced. For instance, the crest elevation should be correctly modelled to ensure similarity in overtopping. It is however important to ensure that the outer envelope of the armour is at the correct level. This may require that the underlayer level is adjusted in the model to accommodate a slightly thicker or thinner armour.

Wave breaking

Typically, scale effects due to wave breaking are not specifically considered in physical stability modelling. They seem to be low for sufficiently sized models. Scale effects are due to the fact that in breaking waves entrained air bubbles are larger in the model because the size is determined by surface tension. Also, the depth of air entrainment will be greater in the model. The total energy budget however, remains in similitude (Le Mehaute, 1990).

Scaling of short duration loads (wave impacts and overtopping)

A so far unresolved problem is the scaling of short duration impact loads (wave impacts and wave downfall pressures due to wave overtopping). Recent investigations have shown that typical scaling techniques such as Froude and Cauchy scaling are unsuccessful in providing a correct representation of peak loads and peak load durations. Short duration peak pressures/loads can be significantly larger than previously documented, particularly when measured with rapidly responding devices at high sampling rates. Since the stability of massive structures is usually not affected by short duration impact loads, simple Froude scaling is likely to provide conservative results. Local pressures/damages may however be more strongly affected by high-intensity short duration peak loads. Since localised damage can result, over time and if not properly repaired, in damage progression, structural integrity failure and ultimate failure of whole structure sections, proper scaling procedures for short duration impact loads require further research. Where possible, model scales and data acquisition rates should be maximised.

A preliminary assessment of short duration impact loads and structural integrity failure can be found in Bullock *et al.* (2004, 2005, 2007), Wolters *et al.* (2004, 2005), Marth *et al.* (2005), Bezuijen *et al.* (2005).

Wave reflection

It is accepted that physical model tests should be conducted at large enough scale to prevent significant model effects on wave reflection. A large model scale should thus be used with armour sized according to geometric scale (Hughes, 1993a) where possible. If deemed necessary for the project at hand, the size of the armour can be increased above the geometric scale to adjust the permeability of the structure. For further information on wave reflection and possible wave trapping between structure and wave generator, refer to chapter 2.

Wave overtopping

Model and scale effects in wave overtopping at model scale are induced by varying slope roughness, structure permeability and by wind effects for example. For smaller armour units and low overtopping volumes ($q < 1$ l/s/m) the combination of model, scale and wind effects can increase. For further information refer to the CLASH report (De Rouck & Geeraerts, 2005).

Model and scale effects

Model and scale effects can distort the modelling results. Model effects are introduced by the model set-up in the laboratory itself, for example by creating artificial boundaries, by inadequate wave damping etc. Scale effects occur when the employed scaling law does not correctly reproduce the physical conditions from prototype at model scale. This can be due to an oversimplification or omission of the governing forces in the physical process.

Scale effects in short-wave hydrodynamic models result primarily from the scaling assumption that gravity is the dominant force balancing the inertial forces. Scaling based on this assumption (Froude scaling) incorrectly scales the other physical forces of viscosity, elasticity, surface tension etc., with the belief that these forces contribute little to the physical processes. Engineers have learned to live with these scale effects and quantify them. Experience and judgement is used to interpret and correct the model results. The non-similitude of viscous forces and surface tension forces in Froude scaled models can lead to scale effects involving wave reflection, wave transmission, wave energy frictional dissipation, and wave breaking dissipation. Therefore it is very important to clearly define the study objective and to recognize important scale effects to allow an intelligent analysis of a particular modelling problem. To prevent model effects, the following provisions should be taken in the layout of the model:

- The model positioning in the wave channel/basin should be such that boundary effects are minimized and that the given wave conditions are achieved over the appropriate test section(s) and produce the appropriately scaled responses.
- Guide walls can be applied where appropriate to control energy spreading/diffraction effects (due to insufficient wave crest length in the model). A minimum

distance between (the end of the) guide walls and the test sections should be guaranteed so that no adverse effects due to the wall are experienced on the test section.

- Wave dampers should be used to prevent that (re-) reflections from model boundaries distort the incident wave conditions at the structure.
- Structure and measurement equipment should be appropriately fixed to prevent that structural oscillations influence the test results.
- The model scale should be as large as possible.

Modelling limits

Modelling limits are defined on the one hand by the (maximum) size of the available modelling facility and on the other hand by the similarity laws (minimum size). The lower limit for the model size is determined for example by the Reynolds number, which must always be large enough to guarantee fully turbulent conditions in the model if these are also found in prototype. Also the Weber number must be large enough to guarantee no influence of surface tension (wave damping) in the model. From these considerations the following model boundary values can be derived, see also Le Mehaute (1976) and Hughes (1993a).

- water depth: >5 cm
- wave height   >2–3 cm (design wave height: >5 cm)
- wave period   realistic wave steepness
- rock diameter   >3–5 mm
- rock armour   >25 mm

Concrete armour unit sizes are usually limited in the model by the available/rentable sizes (often >30 mm). A practical limitation can be that the detection of rocking becomes increasingly difficult for small concrete unit sizes. For the modelling of structural behaviour (modelling of reflections, open/permeable structures, wave pressure distribution) a significantly larger wave height should be aspired to. In each case, physical modelling experience is needed to determine the correct scale and to interpret the measurement results. The following scales are commonly employed for physical breakwater models (e.g. Hudson *et al.*, 1979; Hughes, 1993a; Oumeraci, 1984):

Breakwater stability: 1:5–1:80
                         (typical 2D: 1:30–1:60, 3D: 1:30–1:80)
Forces on solid bodies: 1:10–1:50

3.2.6 Measurement equipment and analysis

The employed instrumentation should provide adequate resolution, be unsusceptible to soiling/dirt and be stable under varying temperatures. Current/wave induced structural oscillations should not affect the output of the instruments. Employed instruments are:

- Wave probes (resistance or capacitance type);
- Directional wave gauges (to determine the wave direction; e.g. composed of a coupled velocity (u, v) and wave gauge (η));
- Velocity meters (e.g. electromagnetic gauges, LDV, acoustic Doppler techniques or simple step/wave gauges);
- Pressure/force sensors (strain gauges, dynamometer);
- Profilers (for damage assessment; used are profilers of wheel type, acoustic type and 3D laser scanners);
- Photographic & video equipment.

Among the specified instrumentation, the photographic equipment is the most widely used and most versatile. It is used throughout the modelling process to document the model set-up, the model operation, the recording of damage and the assessment of wave conditions etc. Depending on the area of instrument application a number of assessment techniques are used:

Damage

Damage is usually assessed using profilers (Figure 3.3) and/or photographic techniques. Digital overlay techniques are employed to assess rock and concrete unit movement. Photographs (taken before and after the test) and videos are also used to assess structural/toe stability. For larger model structures with extensive damage, profilers are often used to assess the damage area. Mechanical profilers are usually

Figure 3.3 Damage assessment using mechanical profiler.

the most robust technique to measure damage, although acoustic, laser and stereo photo techniques are applied successfully. The latter are however typically limited in the measuring area.

Wave run-up and overtopping

Wave run-up is usually assessed using resistance type wave gauges or step-gauges (pressure sensors embedded in the mound slope) and photographic techniques. Wave overtopping is usually assessed by collecting the overtopping water in overtopping trays or tanks and measuring the overtopped water volume or mass. The number of overtopping events can be assessed by a wave gauge at the crest of the breakwater or by continuous water level measurements (volume or mass) within the overtopping tray or tank (Figure 3.4).

Wave loading

Pressures are usually measured using pressure sensors installed within the structure. Force measurements (and moments) are usually conducted by strain gauges or by averaging pressure sensor readings over the given area. For force measurements, it is often necessary to use suspended/independently anchored sensors or sensor arrays (force frames) to produce reliable force estimates. Force sensors are usually only able to resolve global or quasi-static forces. Peak loads arising from wave impacts are not measured with this system, as the force frame cannot respond quickly enough to peaks of very short durations. Information on wave impact loads can generally only be obtained by detailed pressure measurements on the front face of the structure (e.g. caisson). Particular attention needs to be taken when up-lift forces / pressures are measured. Pressure and force sensors can be fragile and are often restricted in their applicability (pressure range, temperature, eigenfrequency range).

Figure 3.4 Measurement of wave overtopping using an overtopping tank.

Velocity

Velocities can be measured by propeller, electromagnetically or using standard wave gauge techniques. Electromagnetic devices (velocity probes) are known to cause problems when used in combination with wave breaking. The entrained air often causes unreliable measurements. In these cases, conventional techniques such as a series of wave gauges, is a better solution to determine the wave celerity.

Velocities within the structure (filter velocity) are usually not assessed. If necessary, protected wave gauges are installed within the structure. Hot-film and hot-wire techniques, LDV and endoscopic PIV are also employed in porous structures. The former probes are very fragile and care must be taken that the probes are regularly cleaned (Hughes, 1993a).

Data acquisition and data processing

Various data acquisition and data processing systems are employed in hydraulic engineering. Many of them have been developed in-house by large modelling facilities. Among those are the DHI Wave Synthesizer (DHI), AUKE (WL), HR software, GEDAP/CHC (SOGREAH). Typical minimum data acquisition rates of 15–25 Hz are employed for wave recording. Large data acquisition rates of 100–1000 Hz (often with a reduced number of channels) are usually used for pressure and load cell recordings. The signals are often filtered during acquisition using low, high or band pass analogue filters to filter out noise or to narrow the frequency range used in the analysis. Digital filters are employed in the analysis of the data. Care must be taken to prevent aliasing, taking into account the eigenfrequency response of both structure and measuring equipment.

3.2.7   Materials

Fluid and structural elements

Fresh water is commonly used in most physical models. In the past also other fluids have been used (e.g. salt water, air, glycerine etc.), but they are today more of academic interest. The varying density from seawater is taken account of in the model (Refer to section on similitude). Structural elements are often constructed of wood, metal, concrete or plastic. Stiff Perspex is used when it is important to get a view of current/wave behaviour inside the structure.

Armour

The rock gradation is usually provided by the designer. Otherwise, rock gradation and established rock sieving curves can be taken from the Rock Manual (CIRIA, CUR, CETMEF, 2007). Before the model testing, the available rock material needs to be checked against the required grading curves using typical sieving/screening techniques or by weighing of the stone material. A shape coefficient for the rock can be taken into account if required. The available rock material should also be checked for proper surface roughness and rock shape. Unrepresentative shapes should be discarded.

If hydraulic stability considerations dictate larger armour units, often surpassing the maximum stone sizes the quarry can produce, artificial concrete armour units are used. These exist in various sizes and geometric configurations. Examples include Dolosse, Tribars, Tetrapods, CORE-LOC™, ACCROPODE™, Xbloc®, hollow cube blocks (SHEDs or COBS), solid cubes and rectangular blocks etc. Fracturing of armour unit limbs should not be an important consideration for the bulky armour units where only shape, surface texture, and the armour unit specific gravity are important similitude considerations (Owen & Briggs, 1985). For slender units (e.g. Dolosse) breakage can occur under prototype conditions. Investigations suggest that for slender armour units, fracturing would most likely occur in units that moved or rocked (Hughes, 1993a). Breakage is usually not modelled in hydraulic laboratories.

For physical models, artificial armour units can usually be rented at varying scales and weights. They can also be produced in-house (refer to Hughes, 1993a for armour unit manufacturing techniques). Armour unit placing is either random or hand-placed according to the licencee's specifications (for concrete armour units such as CORE-LOC™, ACCROPODE™, Xbloc® etc.). The total number of units placed should correspond to prototype conditions and the licencee's specifications. In this context, it is noted that 'keying' of concrete armour units is more difficult underwater where the units are placed 'blindly'. This may require use of different placement patterns/densities in the modelling.

Occasionally, an insufficient amount of model units which can be rented or economic/time constraints, make modelling of the entire structure with concrete armour units inexpedient. In these cases 'dummy' units with similar overall outer shape, size and characteristics of roughness and porosity are introduced into the model. The correct concrete armour units are then reserved for the part of the model exposed to the most severe wave attack or the predominant wave direction. The 'dummy' units are placed in the other areas. If necessary, model tests can be repeated with the correct armour moved along the structure. The width of the tested armour section in the model should be greater than 15–20 rock diameters (D_{n50}) to achieve representative results.

3.2.8   Model operation

A carefully prepared model of a rubble-mound structure is only the first step in conducting a proper model study. Without careful attention to operation of the model, there is a very real possibility that model results will be erroneous. Important points to note are:

- *Inspection*: All scale models should be checked by the engineer in charge before testing. This includes bathymetric elevations, major components of the model such as the crown wall or crest elevations, etc.
- *Shake-down tests*: After the model has been constructed, it should be exposed to lower energy waves to 'shake-down' the structure and allow the armour units to 'nest' into a more compact mound (Hughes, 1993a). Hudson & Davidson (1975) and Tørum et al. (1979) suggest using waves having a height of about 50–60% of the intended test conditions (and the same wave period as the target test condition). The model structure should then be subjected to approximately 1000 waves. This exposure to small waves could be thought of as simulating lower energy waves at the beginning of a storm.

The effects of shake-down tests (structure settling) should always be monitored and documented (e.g. using photographs). Material/armour movements should be quantified. Further settlements of rock material during the tests should be reported as observed.

Procedures for model testing

Procedures for model testing depend on the objectives of the particular model study. Incremental loading procedures are typically used in structure stability testing. A test series of individual tests with increasing intensity (H_{m0}/T_p) are used (typically 3–7 tests) up to design wave conditions. Usually these tests are followed by an overload test for testing reserve stability, typically 110% to 120% of the design wave height. The duration of each individual test is typically about 500–3000 waves. The number of waves should be greater than 1000 to be statistically relevant.

If time permits their execution, repeatability tests are always sensible. Repeatability tests demonstrate the capability of the model to produce similar results under similar forcing conditions. Some scatter can always be expected when repeating the same test conditions. However, time and budget constraints make repeatability tests seldom possible.

Verification of rubble-mound structure models (using a new model) is not often possible. Most studies involve the testing and optimization of a design that has yet to be built, and studies of modifications to existing structures rarely have the necessary prototype data for proper validation. Most model tests rely on the scaling guidance that has been proven correct by several studies in which prototype damage has been re-created in a scale model (Hughes, 1993a). However, many hydrodynamic phenomena studied using short-wave hydrodynamic models do not require verification because the scaling relationships used have been thoroughly tested and proven correct. As long as the model has been carefully scaled and constructed, and it has been determined that laboratory and scale effects are minimal, the engineer can have reasonable confidence that the model is correctly reproducing the hydrodynamic phenomenon (Hughes, 1993a). An example where short-wave hydrodynamic models can benefit by verification of the model to either theory or prototype measurements is studies of existing harbours. It is generally wise to make use of prototype measurements if available. This gives more credibility to the model, and the engineer can be assured that the major hydrodynamic forcing functions have been included in the model.

Extreme waves are usually not addressed in typical physical models. Notable exceptions can be e.g. applications in the offshore sector. In these cases extreme events can be modelled in 2D by wave group focussing, that is the superposition of a wave train with varying wave group velocities which produces an extreme event at a defined point in the wave channel. However, the following reliability aspects need to be addressed during modelling:

- The part of the wave spectrum/point in time where the maximum event will be based (see e.g. Clauss, 2002);
- The storm shape/duration used. A change in response can be expected for other storm durations/storm shapes and repetitive testing;
- Difficulty of correct response: the biggest waves do not necessarily produce the maximum response;

- The amplitude distribution of the wave group used. This determines the wave shape.
- Wave focussing is usually used for deep water cases. In shallower water wave focussing becomes increasingly difficult to predict;
- Since the method of wave group focussing is based on the superposition of linear waves, it is not directly applicable to non-linear effects such as freak waves. The modelling of 'random' freak waves should be done using very long wave flumes with sufficiently long testing periods.

Storm profile modelling

The typically used incremental loading procedures in physical modelling as described in the previous section describes rising storm conditions. However, it is known that the wave period (wave steepness) can vary significantly during a rising/falling storm and that structural failure does not necessarily occur during the rising storm but can occur during a falling storm (Owen & Allsop, 1983). The above indicates that, instead of modelling only the design waves during the rising storm, the modelling of storm profiles can become worthwhile (e.g. for berm breakwaters).

Observations during testing

It is very important that detailed observation accounts are made during testing since they provide a valuable insight into structure specific processes and model effects which can become critical to the ultimate design. The observation account should include observations on overtopping, wave transmission, wave breaking, wave shoaling/diffraction/refraction as well as damage progression, armour rocking and model effects (reflections from model boundaries).

Armour stone can usually tolerate rocking without loss of armour layer integrity. Observation of rock armour rocking can provide qualitative information about the deterioration of the material. If the armour layer consists of slender artificial concrete units, even rocking in place can cause breakage of the units into smaller pieces which are more easily removed by waves. Therefore, if concrete armour units are used in the model, the amount of rocking armour units is usually assessed during testing.

3.3   ANALYSIS PROCEDURES

Various ways of presenting/analysing data will always exist, in fact this should be encouraged in order to facilitate new developments and insights. Therefore, this section only presents some guidelines for fundamental analysis requirements.

Data handling

Usually, test results are presented in dimensional form and prototype values in design studies. Dimensionless analysis of the most important parameters as basis for interpretation of results and their presentation can however give valuable insights into the model behaviour. This is especially useful if compared with other relevant tests or design guidelines or if data is exchanged between varying partners/facilities. Filtering of data after acquisition can facilitate correct data interpretation. For example, short waves can thus be separated from long ones or turbulent fluctuations in the surf zone can be filtered out. Statistic/probabilistic analyses are rarely used in physical modelling practice, partly due to the limited number of tests performed (and generally no repeat tests). However, they become constantly more important in research since many hydraulic (random) processes can be best described in this fashion.

Removal of spurious data

The removal of spurious data is an important prerequisite for an accurate data analysis/interpretation. This includes the removal from the data of the following:

- 'Spikes' due to instrument problems or data acquisition methods;
- Offsets due to the instrument or analogue/digital conversion;
- Slowly varying trends due to instrument drift and changes in water level.

Techniques to remove these 'errors' are given in Mansard & Funke (1988) and Bendat & Piersol (1971). Filtering of data to remove unwanted higher – or lower frequency oscillations is another method to remove spurious data. Using high – and low pass filters short waves can be separated from long waves and turbulent fluctuations in the surf zone can be filtered out (Hamm & Peronnard, 1997).

Damage assessment

Damage occurs to rubble-mound structures when individual armour units on the structure are dislodged or settle. This can lead to further unravelling of the structure or loss of underlayer material. The armour size distribution and its contact with neighbouring units have a great influence on the stability of individual armour units. Therefore, damage should not be expected to be uniformly distributed across the test section of the model, but instead damage will occur in spots (especially for concrete armour units, see Hughes, 1993a; Tørum, et al., 1979).

Two methods are commonly used for quantifying damage in rubble-mound structure models: counting the number of individual armour units that have been dislodged, or determining the volumetric change in areas where armour units have been displaced. The method of counting displaced armour units requires some way of identifying those armour units that have moved. A common technique is to construct the model structure with differently coloured (painted) armour units placed in patterns (Figure 3.5). Dislodged units will then move into a region of a different colour and be easily recognized. The movement can be observed and noted, or more conveniently, video and photographic documentation can be used to record test results.

Quantifying damage by volumetric change requires that pre-test and post-test profiles of the armour slope be measured in a consistent manner for comparison. The test section should be surveyed over a set grid with sufficient resolution to determine profile change with reasonable accuracy. A 'damage' percentage can be defined in a number of different ways. For example, Hudson (1959) defined damage as the percentage of dislodged armour units to the total number of armour units:

$$N_d = \frac{N_{displaced}}{N_{total}} \times 100\% \tag{3.9}$$

Figure 3.5 Damage assessment using sections of painted rock material.

in which $N_{displaced}$ (–) is the number of displaced stones and N_{total} (–) is the total number of stones in that layer (section). The damage percentage is typically calculated for individual sections. Displaced stones are stones which are displaced by more than one unit diameter (D_{n50}).

Damage in terms of displaced units is often given as the relative displacement to the number of units within a specific zone around SWL (CEM, 2006). The reason for limiting damage to a specific zone is that otherwise it would be difficult to compare various structures because the damage would be related to different totals for each structure. Because practically all armour unit movements take place within the levels $\pm H_s$ around SWL, the number of units within this zone is sometimes used as the reference number. However, because this number changes with H_s it is recommended specifying a H_s-value corresponding to a certain damage level (as proposed by Burcharth & Liu 1992) or to use the number of units within the levels SWL $\pm$ n D_{n50}, where n is chosen such that almost all movements take place within these levels (CEM, 2006).

Another method to describe the damage is the use of the N_{OD} number. The N_{OD} number is defined as the number of displaced stones in a strip width of one nominal diameter (D_{n50}).

$$N_{od} = \frac{N_{displ}}{B/D_{n50}} \tag{3.10}$$

Where:

N_{displ} = number of displaced stones (–)
B = width of the test section; determined at the centerline of each section (m)
D_{n50} = nominal stone diameter, exceeded by 50% of the stones (m)

Usually the N_{OD} number gives a better representation of the damage than the damage percentage, since the latter is always dependent on the chosen definition for the related total number of stones. Broderick & Ahrens (1982) introduced a damage level parameter S_d, based on the nominal median armour unit diameter. It is often used to describe the damage to rock slopes:

$$S_d = \frac{A_e}{D_{n50}^{2}}$$ (3.11)

With:
A_e = cross-sectional erosion area (m²)
D_{n50} = nominal median armour unit diameter (m)

Regardless of what method is used to quantify rubble-mound structure damage, it is important to describe the method when reporting results. Furthermore, photographic means and qualitative observations should be used to describe the damage in detail. Damage assessment to concrete armour layers can be simplified by using photographic overlay techniques which are able to distinguish even small armour movements. Damage photos should be taken from the drained model using fixed camera positions whenever possible.

A structure is generally said to have failed once the filter layer is exposed or once a critical damage value is exceeded. For critical values of damage percentage N_d, N_{OD} and damage parameter S_d for varying materials (rock, concrete armour units) and varying armour thicknesses refer to the Rock Manual (CIRIA, CUR, CETMEF, 2007) or CEM (2006). In the case of single-layer armouring of concrete units, usually no damage and only minor rocking is accepted under design conditions (CIRIA, CUR, CETMEF, 2007).

The armour layer should generally be further able to withstand an overload of about 20 percent (design wave height exceeded by 20 percent) without severe damage. Permissible damage levels are also often provided by the designer and manufacturer of the (concrete) armour units. Since critical values can vary largely, an expert appraisal is often recommendable. The structure is often not repaired during individual tests so that an assessment can be made about the cumulative damage after several storm events. Usually the armour layer and toe are reconstructed after a complete test series (e.g. for each wave direction).

Overtopping and wave transmission

The maximum permissible values for wave overtopping and wave transmission depend on structure type and the requirements of the designer. They vary with use of the structure, exposure etc. The Rock Manual (CIRIA, CUR, CETMEF, 2007), the Coastal Engineering Manual (CEM 2006) and the British Standards (BS 6349, 1991) provide possible guideline values. For dikes also the TAW (2002) guidelines can be recommended.

3.4   REPORTING PROCEDURES

Typically the physical modelling reports are made in form of test reports and include:

- Project description & problem definition;
- Project methodology;
- Description of the test facility, model set-up, method of construction, placement of armour units and instrumentation;
- Testing programme;
- Description of test results;
- Conclusions and recommendations.

The Appendix should include:

- Parameter tables of the test results (e.g. damage percentage, overtopping discharge / volume, structural loadings etc.). The recorded wave conditions should also be included (T_p, T_m, $T_{m-1,0}$, H_{m0} or $H_{1/3}$, water levels, wave direction etc.);
- Time series plots for significant data (pressure peaks, typical wave trains);
- Typical energy density spectra;
- Grading curves of materials;
- Surveyed (damage) profiles;
- Drawings of structure cross sections and structure layout in the modelling facility (set-up), including instrument locations and bathymetry contours;
- Photo (video) documentation.

3.5   FUTURE WORK

There are a number of areas in rubble mound breakwater modelling that warrant future research. These are listed below:

- Extension to other breakwater structure types (e.g. impermeable structures, vertical walls, berm breakwaters, composite structure etc.);
- Movable-bed physical models (and scour);
- Structure-foundation interactions;
- Long-wave models;
- Long-term effects;
- Effects of combined waves and currents;
- Probabilistic aspects.

Chapter 4

Sediment dynamics

James Sutherland & Richard Soulsby

Roger Bettes, Iván Cáceres, Rolf Deigaard, Enrico Foti, Joachim Grüne, Luc Hamm,
Jens Kirkegaard, Maarten Kleinhans, Stuart McLelland, Rosaria Musemaci,
Hocine Oumeraci, László Rákóczi, Jan Ribberink, Leo van Rijn, Peter Thorne,
Agustin SanchezArcilla, Joel Sommeria, György Szepessy & B. Mutlu Sumer

4.1 INTRODUCTION

This chapter provides descriptions of essential features of modelling of a broad range
of problems involving many classes of sediment, flow, applications and research topics. The experimental reproduction of hydraulic processes involving sediment dynamics is focussed on the sediments and not the hydrodynamic processes, which are
covered in other chapters. Since a great deal of experimental research on sediment
dynamics remains to be done, laboratory experiments to elucidate process knowledge
(research) as well as physical modelling of site-specific applications (consultancy) will
be covered.

The following items are not discussed within this chapter:

- Soil mechanics and geotechnics, which constitute a major subject area in their
  own right
- Fluidisation/liquefaction of the bed by upward flows or wave action
- Permeable flows

Neither numerical models nor field measurements are discussed.

A thorough and readable account of physical modelling in coastal engineering,
including a substantial chapter on sediment transport models, is given in the book by
Hughes (1993). He offered the opinion: "Understanding sediment transport in coastal
regions is a perplexing challenge that in all likelihood will continue to frustrate coastal
researchers and engineers for generations to come". This opinion could be applied to
non-coastal situations such as rivers with equal force, and is as true today as it was in
1993. The present guideline is not intended to duplicate the detailed (and still generally applicable) accounts given in this and other books. Instead the aim is to summarise key results, and add knowledge derived in recent years, particularly that based on
the experience of HYDRALAB partners.

Most established laboratories that perform physical modelling have their own
particular methods for conducting model studies. Some of the techniques have evolved
over time as a result of experience, while other techniques have been acquired from
other laboratories doing similar types of modelling. In practice physical modelling
procedures in various laboratories vary widely in their methods, such as water and
sediment circulation, waves (regular/irregular, long/short-crested), natural or low

density sediments, scaling methods, methods of measuring sediment concentrations, transport, bottom topography, analysis techniques, etc.

Considering first *research* into sediment processes, a great deal of understanding has been gained in the past 50 years by means of both laboratory and field experiments. However, there are still important gaps in knowledge and the pace of experimentation carries on undiminished into the future. On the other hand, the *consultancy* use of physical models for the prediction of site-specific sediment response has decreased somewhat over the years as the capabilities of numerical models have increased. Yet there are still many classes of application for which a physical model is either the cheapest or the most reliable approach, or is an alternative or complement to a numerical model (composite modelling, the subject of the CoMIBBS JRA), and this is likely to remain true well into the foreseeable future.

The following books, which are widely accepted in the field of physical modelling of sediment dynamics, are used as an underlying basis for this document:

- Yalin, M.S., 1971
- Dalrymple, R.A., 1985
- Hughes, S.A., 1993
- Van Rijn, L.C., 2007

4.2 OBJECTIVES AND APPROACH

The following general approach applies to consultancy applications involving physical modelling of sediment dynamics, and to some extent to research experiments as well. Steps 1 to 10 will normally be considered during the preparation of the proposal, and then refined once the work commences. The layout of this chapter broadly follows the steps of this approach.

1   The objectives of the study must be clearly identified, and a written statement agreed with the client. Misunderstandings at this stage are very difficult to correct later. For example, in a beach study is the plan shape, the cross-shore profile, or the longshore transport rate of primary concern? Establish whether the client wants only the model data, or an interpretation of what the data means. Establish what level of Quality Assurance the client requires.

2   The relevant physical processes must be identified and their approximate magnitudes estimated. The dominant processes must be reproduced in the model, and omission (or non-scaled reproduction) of lesser processes must be justified by consideration of the ratios of omitted to include processes (e.g. ratio of terms in the momentum equation).

3   Decide if the problem can justifiably be treated as having one horizontal dimension (1DH), or whether both horizontal dimensions (2DH) must be modelled. In the former case, a (narrow) current or wave flume will suffice. In the latter case, for rivers either a broad current flume or a full 2DH physical flow model is required, and for coastal and offshore problems a wave basin is required. Using a flume will reduce costs compared with 2DH facilities, and/or a larger number of tests could be performed, but at the expense of omitting cross-flume processes.

4 Consider the scaling issues and the scaling approach to be adopted. Decide whether natural density or low-density sediment will be used and what the minimum scale is that will ensure that non-scaled phenomena have a negligible effect. Determine whether a vertically distorted model should/can be used. Choose the geometric scale (and vertical exaggeration, for distorted models) and calculate the scale of other variables (time, velocity, etc.).

5 Decide what the requirements are for flow generation (where appropriate). This includes considering how the flows will be circulated and the necessary measures for straightening the flows and ensuring that entrance conditions are gradual and turbulence levels are natural.

6 Decide what the requirements are for wave generation (where appropriate):
   a Regular vs irregular waves
   b Long-crested vs short-crested waves
   c Consideration of wave reflections (use of active or passive absorption)
   d Low frequency waves
   e Wave velocity- or acceleration skewness
   f Re-circulation of water (e.g. from longshore currents)? See Chapter 2.

7 Decide how (and if) sediment will be re-circulated.

8 Choose the most appropriate model facility, bearing in mind all the above considerations. Decide whether the scaling benefits of using a large facility (e.g. near-full-scale wave flume, oscillating water tunnel) outweigh the considerable added costs, time and staff resource needed.

9 Decide what needs to be measured and which instruments to use. Consider whether their calibration requirements, accuracy, data logging requirements and data storage.

10 Plan a test series. Leave adequate time for calibrations and preliminary tests and for turn-around between tests.

11 Estimate the costs in conjunction with planning the test series. This is best done as a fixed sum for commissioning (and de-commissioning) the facility including preliminary tests and calibrations, plus a unit cost per test. Estimate the time necessary to complete the test series, allowing 10–20% contingency time for breakdowns etc. Consult with the client to ensure that cost and time are in line with his expectations.

12 Design the model, including moulding of the bed, construction of structures, placement of rock armour or scour protection.

13 Perform the calibrations of the instruments and the current and/or wave generation facilities. Perform preliminary tests to establish the best routine for the main test series.

14 Perform the test series. Examine the first few tests particularly carefully, analysing the data as far as possible, and noting the time taken per test. Adapt the procedures and test series if necessary. Log the data in an organised manner.

15 Keep detailed notes in a dedicated log-book (one book for the whole study, not separate ones kept by different individuals). Ensure that everything is recorded, including sketches where useful. Remember to record water temperature, especially if suspended sediments are involved. Don't trust to memory, or assume that something is too obvious to require a note.

16 Analyse the data, preferably as the test series proceeds.

17   Interpret the data, including conversion of model results to prototype scale.
18   Establish the sources and magnitudes of errors, and quote these together with the interpreted results.
19   Write a report on the study.
20   Archive the data and paperwork in a way that will be retrievable over the number of years required by the client or by QA requirements.

4.3   GENERAL PRINCIPLES

A large number of interacting physical processes can be operating in the prototype situation. These must be considered, and where possible reproduced, in the physical model. Some or all of the following should be considered, depending on the type of problem being tackled.

- Type of forcing: current, tidal, waves, wind, buoyancy, Coriolis;
- Type of sediment: cohesive, noncohesive, mud, sand, gravel, mixtures;
- Mode of transport: threshold of motion, bedload transport, suspended transport;
- Bedforms: ripples, dunes, megaripples, sand waves;
- Waves;
  - Long-crested, short-crested;
  - Asymmetry, skewness, low-frequency waves;
  - Longshore currents;
  - Undertow, rip currents, rip cells;
  - Boundary-layer streaming, mid-water return flows;
- Longshore drift of sediment;
- Breaker bar formation;
- Permeability;
  - Beaches (sand, shingle, both);
- Soil mechanics/geotechnics;
- Fluidisation/liquefaction of the bed;
- Biological activity.

Some of these are discussed in further detail below. A fuller understanding of the principles and processes involved, especially in the prototype situation, can be gained from standard texts on sediment transport such as Fredsøe & Deigaard (1992), Nielsen (1992), Van Rijn (1993, 1998) and Soulsby (1997) (see Bibliography).

4.3.1   Multi-directional versus uni-directional waves

In deep water the sea-states generated by storms are multi-directional. As these waves approach the coast they shoal and refract and the directional spreading of the sea-states reduces. By the time waves break near the shoreline they are commonly quite long crested and it is common practice to represent the sea-state in a physical model as a uni-directional irregular series of waves. This is a better approximation inshore than offshore.

In recent years there has been an increase in the availability of multi-directional wave paddles and so there is less need to make such an approximation in cases where the incident seastate can reliably be shown to be multi-directional (See Chapter 2).

4.3.2 Wave skewness and asymmetry

In deep water, the time series of surface elevations at a point is commonly assumed to be Gaussian, with a Rayleigh distribution of wave heights (Chapter 2). As waves enter shallow water, their shapes evolve from sinusoidal to peaky (like a Stokes wave) with sharp wave crests separated by broad, flat wave troughs. These waves have positive skewness, but near zero asymmetry (of their surface elevation). It has been hypothesized that the larger onshore velocities under the peaked wave crests transport more sediment than the offshore velocities under the troughs.

As waves continue to shoal and break, they evolve from profiles with sharp peaks to asymmetrical, pitched-forward shapes with steep front faces. Broken, sawtooth waves have positive asymmetry, but near zero skewness (of their surface elevation). Water rapidly accelerates under the steep wave front, producing high onshore velocities, followed by smaller decelerations under the gently sloping rear of the wave. Large accelerations generate strong horizontal pressure gradients that act on the sediment. Although the precise mechanisms are not fully understood, it has been hypothesized that if accelerations increase the amount of sediment in motion, there will be more shoreward than seaward transport under pitched-forward, asymmetric waves.

It is therefore important that the correct skewness and asymmetry be generated in the laboratory, in order that the correct form of sediment transport may be generated. This can best be done by generating waves in deeper water and giving them sufficient time and space to shoal naturally before the test section with sediment is reached.

4.3.3 Critical bed-shear stress

Particle movement will occur when the instantaneous fluid force on a particle is just larger than the instantaneous resisting force related to the submerged particle weight and the friction coefficient. The degree of exposure of a grain with respect to the surrounding grains (hiding of smaller particles resting or moving between the larger particles) can affect the forces at initiation of motion. In-bed pressure gradients can lift small embedded particles out of the bed. Cohesive forces are important when the bed consists of appreciable amounts of clay and silt particles.

Initiation of motion in steady flow occurs when the dimensionless bed-shear stress (Shields parameter) $\theta = \tau/[(\rho_s-\rho_f)gD_{50}]$ is larger than a threshold value $\theta_{cr} = \tau_{cr}/[(\rho_s-\rho_f)gD_{50}]$. Thus: $\theta > \theta_{cr}$, with τ_{cr} = critical bed-shear stress for movement of noncohesive sediment, ρ_s = sediment density, ρ_f = fluid density, D_{50} = median sediment diameter and g = gravitational acceleration.

The θ_{cr}-factor depends on the hydraulic conditions near the bed, the particle shape and the particle position relative to the other particles. The hydraulic conditions near the bed can be expressed by either the grain Reynolds number $Re* = u*D_g/v_k$ or by a dimensionless particle size $D* = D_{50}[(s-1)g/v_k^2]^{1/3}$, $s = \rho_s/\rho$ = relative density, $u*$ = friction velocity and v_k = kinematic viscosity. The Reynolds number and the

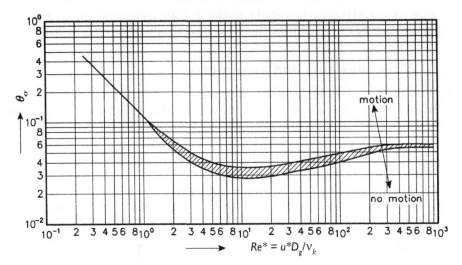

Figure 4.1 Initiation of motion according to Shields (1936), as presented by Van Rijn (1993).

Shields parameter should ideally be the same in model and prototype, but this can usually only be accomplished in facilities which represent the processes at full (or nearly-full) scale, otherwise a compromise must be made.

Many experiments have been performed to detemine the θ_{cr}-values as a function of Re^* or D^*. The experimental results of Shields (1936) relating to a flat bed surface are most widely used to represent the critical conditions for initiation of motion (Figure 4.1). The Shields-curve represents a critical stage at which only a minor part (say 1% to 10%) of the bed surface is moving (sliding, rolling and colliding) along the bed. The scaling requirements to meet these conditions are discussed in section 4.4.

4.3.4   Bed forms in steady flows

The bed of a river, estuary or sea might under different circumstances be either flat or exhibit irregularities (bed forms). The bed form regimes for steady flow over a sand bed can be classified into (see Figure 4.2):

- Lower transport regime with flat bed, ribbons and ridges, ripples, dunes and bars,
- Transitional regime with washed-out dunes and sand waves,
- Upper transport regime with flat mobile bed and sand waves (anti-dunes).

A flat immobile bed may be observed just before the onset of particle motion, while a flat mobile bed will be present just beyond the onset of motion. The bed surface before the onset of motion may also be covered with relict bed forms generated during preceding stages with larger velocities.

Small-scale ribbon and ridge type bed forms parallel to the main flow direction have been observed in laboratory flumes and small natural channels, especially in

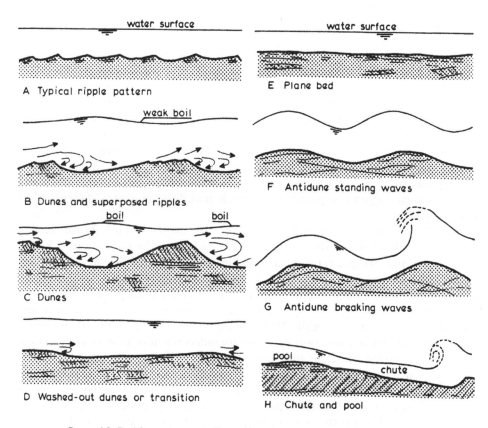

Figure 4.2 Bed forms in steady flows (rivers), as presented by Van Rijn (1993).

the case of fine sediments ($D_{50} < 0.1$ mm). Longitudinal bedforms aligned parallel to the mean direction of flow can form in both homogenous and heterogeneous sediments (e.g. Karcz, 1973). These bedforms usually take the form of periodic, lateral variations in bed texture (sediment stripes) and/or bed topography (sediment ridges). Sediment stripes usually form under low shear stress conditions and may replace lower stage plane beds with sediment stripes in bimodal sediments (e.g. McLelland *et al.*, 1999). The formation of sediment stripes/ridges is usually attributed to the lateral transport of sediment particles by a series of streamwise oriented vortices. Adjacent vortices with alternate directions of rotation transport sediment away from regions of flow divergence at the bed forming troughs of fine sediment, and towards regions of flow convergence forming ridges or stripes of coarser bed sediments. In narrow open channels with width, *b*, to depth, *h*, ratios of *b/h* < 5 secondary currents generated at the channel corners extend to the central region of the channel whereas in wider shallower channels (*b/h* > 5) the central section of the channel is not influenced by these corner-generated flow (Nezu & Rodi, 1985). However, in the absence of corner generated secondary flows, spanwise variations in bed topography or boundary roughness may also be sufficient to induce secondary flows that lead to lateral sediment transport processes. For experiments undertaken at low shear stresses where the sediment transport rates are low it is therefore important to consider the width to depth ratio of the channel since this will affect the resulting flow structure and development of longitudinal bedforms.

When the bed form crest is perpendicular (transverse) to the main flow direction, the bed forms are called transverse bed forms, such as ripples, dunes and anti-dunes. Ripples have a wave length which is practically independent of the water depth, and may be much smaller than the water depth, whereas dunes have a length scale much larger than the water depth. Ripples and dunes travel downstream by erosion at the upstream face (stoss-side) and deposition at the downstream face (lee-side). Bed forms with their crest parallel to the flow are called longitudinal bed forms such as ribbons and ridges. In the literature, various bed-form classification methods for sand beds are presented. The types of bed forms are described in terms of basic parameters (Froude number, suspension parameter, particle mobility parameter, dimensionless particle diameter).

When the velocities are somewhat larger (by 10%–20%) than the critical velocity for initiation of motion and the median particle size is smaller than about 0.5 mm, small (mini) ripples are generated at the bed surface. Ripples that are developed during this stage remain small with a ripple length much smaller than the water depth. Current ripples have an asymmetric profile with a relatively steep downstream face (lee-side) and a relatively gentle upstream face (stoss-side). Their height is typically 2 cm and their wavelength typically 20 cm. Some evidence suggests that their dimensions depend only weakly on grain size, and are independent of flow velocities if the experiment is allowed sufficient time for the ripples to fully reach equilibrium. As the velocities near the bed become larger, the ripples become more irregular in shape, height and spacing yielding strongly three-dimensional ripples. The largest ripples may have a length up to the water depth and are commonly called mega-ripples.

Another typical bed form type of the lower regime is the dune-type bed form. Dunes have an asymmetrical (triangular) profile with a rather steep lee-side and a gentle stoss-side. The length of the dunes is strongly related to the water depth (h) with values in the range of 3 to 15 h.

The largest bed forms in the lower regime are sand bars, which usually are generated in areas with relatively large transverse flow components (bends, confluences, expansions). Alternate bars are features with their crests near alternate banks of the river. Transverse bars are diagonal shoals of triangular-shaped plan along the bed. One side may be attached to the channel bank. These type of bars generally are generated in steep slope channels with a large width-depth ratio.

Bed forms generated at low velocities are washed out at high velocities. Flume experiments with sediment material of about 0.45 mm show that the transition from the lower to the upper regime is effected by an increase of the bed form length and a simultaneous decrease of the bed form height. Ultimately, relatively long and smooth sand waves with a roughness equal to the grain roughness are generated.

In the transition regime the sediment particles will be transported mainly in suspension. This will have a strong effect on the bed form shape. The bed forms will become more symmetrical with relatively gentle lee-side slopes. Flow separation will occur less frequently and the effective bed roughness will approach that of a plane bed.

In the supercritical upper regime the bed form types will be plane bed and/or anti-dunes (which travel *upstream*) with a length scale of about 10 times the water depth. Bedforms can be washed out at any Shields number under large Froude numbers ($Fr = U_c/(gh)^{1/2} > 0.84$), and by large Shields parameters ($\theta = \tau/[(\rho_s-\rho)gD_{50}] >~0.8$) for subcritical flows.

By including a description of the pertinent physical processes, linear stability analysis can explain the formation of bed forms on an initially plane sediment bed, and the theory can account for the conditions under which the bed will remain plane. The linear stability analysis starts from a model for the flow and sediment transport over a plane bed. The bed is given a small amplitude harmonic perturbation and the perturbed flow and sediment transport field is described from the linearized equations. The sediment transport field may modify the bed perturbations causing migration and growth or decay. The analysis can be made for all possible perturbations (wave length and orientation relative to the mean flow). If all perturbations are found to decay the bed will remain plane, while if some perturbations tend to grow the bed is unstable and bed forms will emerge. The perturbation with the fastest growth rate will determine the initial dimensions of the emerging bed forms, but the shape and length scale may be modified as the bed forms grow and the assumption of the amplitude being small is no longer valid. The stability analysis covers bed forms of practically all length scales from ripples, dunes, anti-dunes, alternate bars to braid bars. Stability theories for bed forms in rivers were developed in the 1960s and 1970s and were applied to describe the formation of analogous marine bed forms (sand waves, linear tidal sand banks) in the 1990s.

4.3.5   Bed forms under waves

In coastal regions the non-linear interactions between the flow and the non-cohesive sandy bottom may induce significant modifications in the bottom morphology leading to the appearance of several types of bedforms of different sizes and shapes, from small scale (ripples) to medium (dunes, megaripples) and large scale (sand waves). The bed generally is dominated by small-scale and large-scale ripples and plane beds depending on the wave conditions. The nature of the sea bed (either ripples or plane bed) has a fundamental role in the transport of sediments by waves and currents. The configuration of the sea bed controls the near-bed velocity profile, the shear stresses and the turbulence and, thereby, the mixing and transport of the sediment particles. For example, the presence of ripples reduces the near bed velocities, but it enhances the bed-shear stresses, turbulence and the entrainment of sediment particles resulting in larger overall suspension levels.

Several types of bed forms can be identified, depending on the type of wave-current motion and the bed material composition. Focussing on fine sand in the range of 0.1 to 0.3 mm, there is a sequence starting with the generation of rolling grain ripples, to vortex ripples and, finally, to upper plane bed with sheet flow for increasing bed-shear. Rolling grain ripples are low relief ripples that are formed just beyond the stage of initiation of motion. These ripples are transformed into more pronounced vortex ripples due to the generation of sediment-laden vortices formed in the lee of the ripple crests under increasing wave motion. These ripples are characterized by a height of the order of a few centimetres and a length of the order of some tens of centimetres, comparable to the amplitude of fluid oscillation close to the bed and increasing with the sediment coarseness (Figure 4.3b). Steep ripples are formed by relatively low waves in relatively deep water (i.e. typical offshore conditions). Such ripples tend to be long crested (two-dimensional), and they have steepness (height/length) greater than about 0.15.

In the lab, the appearance of ripples is controlled by the value of the orbital amplitude, the sediment grain size, and the wave characteristics. In particular, when the bottom orbital amplitude is not big enough to mobilize the sand bed, the bottom remains flat. When the velocity increases and it exceeds about 1.2 times the critical value above which the sediments start to move, ripples appear. Finally, when the orbital velocity generates such shear stresses to mobilize the sandy bottom entirely, bedforms reach a maximum height and then decline and eventually disappear resulting in plane bed sheet flow characterised by a thin layer of large sediment concentrations.

When making experiments to investigate the formation of bed-forms, the bed should be carefully smoothed out without compacting the sand, as any sand grain can act as a disturbance to ripple evolution. The evolution of the rippled bed, starting from a flat bed, tends asymptotically towards equilibrium with a time scale of some hundreds of wave periods. At equilibrium, bedforms are characterised by a wavelength depending on the flow and the sediment Reynolds numbers and migrate at a rate depending on the same non-dimensional parameters (Faraci & Foti, 2002). Ripple crests are oriented perpendicular to the wave propagation direction. However in the presence of superimposed orthogonal currents, a waviness of the crest line occur, in the range $0.2 < U_c/U_w < 0.6$ (U_c being the current velocity and U_w the orbital velocity), while for stronger currents ripple crest lines become irregular (Andersen & Faraci, 2003).

Dunes and megaripples have been sometimes considered as a single category, as their characteristics show very small differences. They are characterized by a height of 0.1–0.5 m and a wavelength of 1–5 m. Megaripples tend to become aligned in such a way that the sediment transport normal to the bedform crest is maximized and therefore they migrate (Gallagher *et al.*, 1998). Even in this case most of the studies refer to field conditions.

Large scale bedforms are characterized by a median length of hundreds of meters and an amplitude of about 5 m (Hulscher, 1996) and their formation is related to the presence of steady recirculating cells generated by the interaction of the oscillatory tidal flow and the bottom waviness (Figure 4.3a). Their reproduction at reduced scale in laboratory is difficult due to the matching of hydrodynamic forcing and morphodynamic response and therefore most of the studies are referred to field conditions.

(a)   (b)

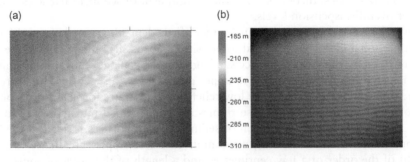

Figure 4.3 (a) sand waves in the Messina Strait (b) ripple bed formed at University of Catania Hydraulics Laboratory.

4.3.6   Bed roughness

Nikuradse (1932) introduced the concept of an equivalent or effective sand roughness height (k_s) to simulate the roughness of arbitrary roughness elements of the bottom boundary. In the case of a movable bed consisting of sediments the effective bed roughness mainly consists of grain roughness generated by skin friction forces and of form roughness generated by pressure forces acting on the bed forms. Similarly, a grain-related bed-shear stress and a form-related bed-shear stress can be defined. The effective bed roughness for a given bed material size is not constant but depends on the flow conditions.

There is an essential difference between the sediment transport processes that occur above rippled and plane sand beds. The roughness (k_s) of a rippled bed is equivalent to about 3–4 ripple heights. Beneath steeper waves in shallower water (for example, at edge of the surf zone), the ripples start to become washed out; their steepness decreases, causing k_s to decrease, and they tend to become shorter crested (transitional 2D-3D profiles). Finally, beneath very steep and breaking waves, the ripples are washed away completely, the bed becomes plane ('sheet flow' regime), and k_s decreases still further (scaling on the sand grain diameter). However, the roughness during sheet flow is larger than it would be due to grain roughness alone as a result of momentum transfer by saltiating grains, resulting in a roughness of typically $k_s \sim 100\, D_{50}$.

Correct modelling of the effective roughness in the hydraulically rough regime requires that for steady currents the ratio of water depth h and effective bed roughness k_s, and for waves the ratio of wave orbital excursion A to k_s, is the same in the scale model and in the prototype. Generally, this cannot be achieved in a movable bed model. Corrections are necessary to deal with this.

4.3.7   Rivers

The amount of water conveyed through a river section during a year is known as the hydrograph, which essentially is a plot of discharge versus time. Changes in discharge cause changes in water level of the river channel. At very high discharges (greater than bank-full discharge) a river may overflow its banks on to the adjacent plains (flood plain). The flood plains are generally narrow in the upper reaches and wide in the lower reaches. In heavily populated areas the flood plain is usually intersected (reduced in width) by artificial banks (dykes). Natural rivers often show meandering patterns within the flood plain areas, while controlled rivers are confined within the fixed banks (dykes). In other cases, natural rivers may have single or braided channels. Sediments transported by a river are in the range from very fine clay minerals to large-scale cobbles and boulders. The sediment accumulated in the river bed is the alluvial part of the sediment load. The fine sediments not present in the bed are carried in suspension as wash load. The total sediment outflow from a catchment area is called the sediment yield of that area.

Scale models are specifically useful for the design of river improvement works and hydraulic structures related to three-dimensional and time-dependent phenomena. Model calibration should be done by using data of the known actual situation.

4.3.8   Coastal areas

Physical scale models have been used frequently to study coastal engineering problems. Scaling laws for movable bed models are well established (Noda, 1972; Kamphuis, 1972, 1982; Hughes, 1993), but the errors due to scale effects are less well understood.

Bed load and suspended load

Sand transport in a coastal environment generally occurs under the combined influence of a variety of hydrodynamic processes such as winds, waves and currents. Figure 4.4 gives a schematic representation of sand transport mechanisms along a cross-shore bed profile of a straight sandy coast.

Sand can be transported by wind-, wave-, tide- and density-driven currents, by the oscillatory wave motion itself due to the transformation of short waves under the influence of decreasing water depth (wave skewness), or by a combination of currents and short waves. The waves generally act as sediment stirring agents; the sediments are then transported by the mean current. Low-frequency waves interacting with short waves may also contribute to the sediment transport process.

In deeper water outside the breaker (surf) zone the transport process is generally concentrated in a layer close to the sea bed; bed-load transport and suspended transport may be equally important. Bed load transport dominates in areas where the mean currents are relatively weak in comparison with the wave motion (small ratio

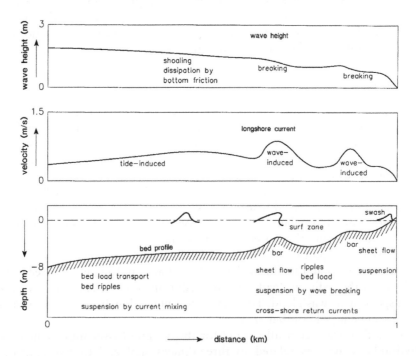

Figure 4.4 Sand transport mechanisms along cross-shore profile. Top: cross-shore distribution of wave heights, Middle: cross-shore distribution of longshore current, Bottom: sand transport processes along cross-shore profile.

of depth-averaged velocity and peak orbital velocity). Suspension of sediments can be caused by ripple-related vortices. Suspended load may dominate even in the absence of a current, when the suspended sediment is concentrated in the thin turbulent near-bed wave boundary-layer. The suspended load transport becomes increasingly important with increasing strength of the tide- and wind-driven mean current, due to the turbulence-related mixing capacity of the mean current (shearing in boundary layer). By this mechanism the sediments are mixed up from the bed-load layer to the upper layers of the flow.

In the surf zone of sandy beaches the transport is generally dominated by waves through wave breaking and the associated wave-induced currents in the longshore and cross-shore directions. The longshore transport in the surf zone is also known as the longshore drift. The breaking process together with the near-bed wave-induced oscillatory water motion can bring relatively large quantities of sand into suspension (stirring) which is then transported as suspended load by net (wave-cycle averaged) currents such as tide-, wind- and density (salinity)-driven currents. The concentrations are generally maximum near the plunging point and decrease sharply on both sides of this location.

Gravel and shingle beaches

Gravel beds and shingle beds are also amenable to physical model studies, and, indeed, this is often the best way of determining impacts of projects (e.g. blocking of longshore transport by jetties, training walls, etc.) and the effectiveness of control structures (e.g. groynes, sea walls). The transport is almost entirely by bedload, which simplifies the scaling arguments. Permeability is an important process, particularly in the achieving the steepness associated with many shingle beaches, and the formation of berms. The use of low-density sediments can be helpful in achieving the necessary permeability at the same time as achieving similitude of transport rates. Experiments can have relatively short durations because the beaches adapt quickly to the imposed wave conditions, especially if low-density sediments are used. However, as noted in section 4.4.3, low-density sediments should only be used with a full appreciation of the theoretical and practical problems associated with them.

Correct modelling of bed load transport and suspended load transport requires that the dimensionless parameters controlling bed and suspended load transport processes are the same in the scale model and in the prototype. These controlling parameters are:

- Dimensionless particle size $D* = D_{50}[(s-1)g/v_k^2]^{1/3}$;
- Shields parameter $\theta-\theta_{cr}$;
- Current suspension number $Z = w_s/u*$ with w_s = fall velocity, $u*$ = bed-shear velocity; and
- Wave suspension number (or Dean fall speed parameter) H/w_sT with H, T = wave height, period.

When planning the duration of experiments with a sand bed, it should be borne in mind that ripples can take a long time to become fully developed, especially under flow velocities that are not much above threshold. Some exploratory experiments might help to determine the necessary durations.

Cross-shore morphology

Although, coastal morphology is three-dimensional, a two-dimensional cross-shore modelling approach (wave flume) is meaningful for conditions with a straight uniform coastline and parallel depth contours. However, along a uniform beach longshore features like beach cusps, rhythmic bars, rip channels and other phenomena may also occur and cannot be rigorously ignored. To a certain extent the longshore variability can be eliminated by spatial averaging (over a typical longshore scale; say order of 1 km) yielding longshore-averaged profiles.

Various subzones can be distinguished in cross-shore direction:

- Dunes (fore and back dunes):the supratidal zone landward of the toe of the foredune;
- Beach (fore and back beach):the intertidal zone between the toe of the foredune and the low tide line; berms may be present on the back beach; swash bars may be present on the forebeach;
- Shoreface (upper, middle, lower):the subtidal zone consisting of:
  - *upper shoreface* with breaking waves and breaker bars (surf zone),
  - *middle shoreface* with shoaling waves and,
  - *lower shoreface* merging into the shelf zone.

Ripples are usually the dominant type of small-scale bed forms in the upper, middle and lower shoreface zone. As the current component gains in strength, the ripples become more asymmetrical and larger in height and length. Generally, the ripples in the upper and middle shoreface are washed out during storm events and moderate waves to leave a flat bed in the surf and swash zones.

Various descriptive classifications are used to identify the type of cross-shore profile (including the beach):

- Equilibrium and non-equilibrium profiles,
- Barred and non-barred profiles,
- Dissipative and reflective profiles.

Equilibrium and non-equilibrium profiles

An equilibrium profile is an idealized profile that has adjusted to the sediment, wave, and water level fluctuations at the site of interest. A cross-shore profile is defined to be in a state of (dynamic) equilibrium if the volume of sand accumulated under the profile and a chosen horizontal datum is constant in time. Ideally, an equilibrium profile represents a bed profile generated under constant wave energy conditions with a constant water level in the absence of longshore transport gradients for such a long time that a stable profile is obtained. Laboratory flume experiments are often designed to simulate the formation of such profiles.

Non-equilibrium profiles caused by the presence of longshore and cross-shore transport gradients, can be classified as eroding/accreting, steepening/flattening and shifting. Eroding profiles can typically be observed downdrift of man-made structures normal to the coastline such as long breakwaters, and in the vicinity of sediment-importing tidal inlets. Erosion of the upper shoreface may lead to profile steepening, if the position of the shoreline is maintained by beach nourishment (with or without a sea wall). Accreting (and flattening) profiles can be observed in the updrift zone of headlands or structures normal to the coastline. Often, accreting profiles have an upward convex shape in the surf zone. These profiles can be found in areas with abundant sediment supply: in front of forelands, at distal ends of spits, along delta coasts and along outer edges of coastal plains. Shifting profiles can be associated with the effect of sea level rise. The nearshore profile maintains its shape while it is continuously adjusting to the rising sea level.

Barred and non-barred profiles

A profile exhibiting a monotonically sloping bed surface is termed a non-barred profile. Under certain conditions bed features are generated along the profile, which are known as bar-trough systems with their crests more or less parallel or sometimes at a very small angle to the coastline. Single or multiple bar systems may be generated. Bars may be either intertidal or subtidal features.

Storm waves attacking a monotonically sloping profile, will often produce offshore transport resulting in the generation of a single submerged bar. After the storm event during periods with low wave energy the sediments are transported in onshore direction by asymmetric shoaling (non-breaking) waves and a berm type profile (or step type profile) is gradually formed and the beach will eventually be restored. The berm on the beach can also be interpreted as a swash bar formed by wave runup processes and welded to the beach.

Dissipative and reflective profiles

Focussing on the hydrodynamic processes, the reflective and dissipative profiles can be distinguished.

Features of reflective profiles are:

* Most of the incident wave energy is reflected on a relatively steep beach;
* Featureless profile associated with erosion (although sometimes a reconstructive swell can make them accretive);
* Predominantly occurring on ocean-fronted beaches composed of relatively coarse material, on beaches in eroding areas and in deeply indented compartments.

Features of dissipative profiles are:

* Most of the wave energy is dissipated by breaking of waves in a wide surf zone consisting of a concave upward nearshore zone and a flat shallow beach;
* Profile exhibits many features such as ridges and runnels, swash bars, longshore breaker bars and troughs, rip channels;
* Predominantly occurring on exposed high-energy beaches in deltaic regions composed of medium/fine sands (open marginal sea coasts).

With regards to modelling, a clear distinction should be made between quasi-equilibrium profiles and time-developing profiles. The distinction between the processes occurring on (steep) reflective beaches and (shallow-gradient) dissipative beaches is important to recognise in experiment design, because wave flumes are usually too short to accommodate a shallow-gradient (e.g. 1:50–1:100) beach. Hence it is easier to simulate reflective than dissipative beaches in flumes. Distorted models (vertical scale different to horizontal scale) are sometimes used, although the breaking processes and run-up may not be correct. These scale effects can be compensated to some extent by modifying the grain-size scale, but this will inevitably be a compromise and some other aspects might not be adequately represented.

Longshore morphology

The morphology of both small-scale alongshore beach forms (beach cusps, breaker bars, rip channels, sand waves) and large-scale shoreline forms (beach plains, headlands, bays, tombolos, spits, forelands) is related to depositional and erosional processes acting along the coast under the influence of waves and currents.

Generally, three main groups of shoreline configuration are distinguished:

- Low-lying (low-gradient) relatively straight coastal plains;
- Irregular indented or embayed coasts;
- Protruding steep-gradient cliff-type or headland-type coasts.

The evolution of coastlines is related to two basic groups of morphological processes, those causing coastal straightening and those causing coastal irregularity. Further details of these processes are given by Van Rijn (1998). With regards to modelling distorted models should not be used for scale models of coastal plan studies as the refraction patterns will not be correct. It is not clear how these scale effects can be corrected for.

Beach drift and permeability

A characteristic feature in the swash zone during low-energy conditions is the zig-zag movement of the sediment particles which is also known as beach drifting. In case of oblique wave incidence, the swash will run up the beach in the direction of wave propagation, but the backwash will move down the steepest slope under the influence of gravity. This latter movement usually is at a right-angle to the shore. Sediment particles being moved by the swash and backwash will follow a zig-zag pattern along the shore, parallel to the front of the breaking waves.

The water carried in the uprush percolates partly through the sediment surface down to the water table, which typically lies a little above the instantaneous tidal level. This percolation reduces the volume of downwash, particularly for coarse gravel/shingle beaches, causing the sand carried up to be deposited partly on the beach face. This build-up of the beach continues during low-energy conditions. During high-energy conditions with breaking waves (storm cycles), the beach and dune zones of the coast are heavily attacked by the incoming waves, resulting in a raised water table and erosion of the beach (Nielsen, 1999).

4.4   SEDIMENTS AND SCALING LAWS

4.4.1   Properties of sediment

Types of sediments

Most sediments are derived from the weathering of rock on land by the action of wind, water and ice under the influence of temperature, pressure and chemical reactions. Such particles range in size from large boulders to colloidal size fragments and vary in shape from rounded to angular. They also vary in specific gravity and mineral composition (quartz and clay minerals). Clay particles have a sheet-like structure, and cohere together due to electrostatic forces (cohesive forces) in a saline environment. Consequently, there is a fundamental difference in sedimentary behaviour between sand and clay minerals.

Sediment particles between 0.062 and 2 mm are usually referred to as sand particles. Based on mineral and chemical composition, three types of sand can be distinguished:

- Silicate sands (quartz and feldspar);
- Carbonate sands (calcite and aragonite, originating from shell and coral fragments);
- Gypsum sands (crystal forms of gypsum, moderately soluble in water).

Many beaches and shorefaces consist of quartz sand. Some beaches consist of materials from volcanic sources, such as the black sand beaches of Hawaii and Iceland. Beaches may also consist of biogenic materials like carbonates (white sands) from coral reef sources. Shells and shell fragments are found on many beaches, especially in the tropics. They can usually be treated for physical modelling purposes as being hydraulically similar to the mineral grains they are found with (i.e. only the mineral grains are considered in the physical model and the shell fragments are assumed to behave similarly).

In some places in this chapter the grain-size is written in terms of D_{50} (the median grain diameter by mass). Where appropriate, other grain percentiles such as D_{90} (the size for which 90% of the sediment by mass is finer) are specified. The most important distinction for modelling purposes is between non-cohesive sediments and cohesive sediments.

Non-cohesive sediments

Fluvial bedload transport

Physical modelling of sediment transport using gravel-sized sediments is important for improving and validating theoretical models of bedload transport rates where field data may be limited. Gravel-sized sediments have been used to model transport rates for a single characteristic grain size such as the median grain diameter (D_{50}) and to investigate fractional transport rates for different sizes within a grain size distribution. The grain size distribution of the sediment used is important since different grain sizes will have different relative mobilities for a given set of hydraulic conditions (flow velocity, water depth and slope). Whilst grain size distributions can be chosen to replicate natural environments with appropriate information from a prototype environment, it is more difficult to reproduce the boundary conditions such as upstream sediment supply and bed surface texture and structure. Upstream sediment supply may be simulated by controlling the sediment feed rate and composition, but this may be limited by a lack of field data. Reproducing natural bed surface textures and structure is more complex since the nature of bed surface deposit will depend upon the antecedent hydraulic regime and the characteristics of the sediment supplied to a reach. Natural bed surface textures and structures cannot easily be reproduced and the nature of the initial bed surface may affect the sediment transport rate. In natural streams, rivers and offshore areas, the bed may be *armoured*, in which the finer grains have been winnowed away leaving a surface layer in which the gaps between coarse grains a filled with successively finer grains, sometimes with biological adhesion as well. Armoured beds are only mobile under extreme flow conditions.

Mixtures

Graded sediments or mixtures often characterise river and coastal sandy beds. However, as pointed out by Parker (1991), the problem of mixtures was in the past mainly faced by generalising results obtained on uniform materials. However, as coastal sediment typically has a wide range of grain size, in recent decades laboratory studies concerning grain sorting have assumed a greater relevance. Such relevance is even more crucial when problems related to bottom morphology are concerned. Indeed laboratory experiments carried out with sediment mixtures have shown that when the bedforms are analysed, a selective sediment transport may take place, enhanced by the presence of sedimentary structures. In the case of ripples, for example, as shown in Figure 4.5, the coarser fraction tends to accumulate along the ripple crests and the finer fraction in the troughs, even though large fluctuations of the grain size distribution within the wave cycle occur (Foti & Blondeaux, 1995). As a result, mixed sediments tend to stabilize the bottom, thus delaying the appearance of bedforms. Furthermore, bedforms which appear in the presence of graded sediments are characterized by longer wavelengths than those produced on uniform sand characterised by the same median diameter under the same wave conditions.

Graded sediments influence sediment transport significantly: larger particles have a smaller mobility than smaller ones, leading to a selective transport process of size fractions in a mixture. In order to consider such a mechanism, the so called hiding function, introduced by Einstein (1950) and Hirano (1971) among others, has to be taken into account. Moreover during sediment transport due to currents a vertical sorting takes place in the upper layers of the sandy bed, affecting also the concentration profiles: within such phenomenon, fine fractions seem to be driven by diffusive processes, while coarse sediments seem to be controlled by advective processes. Laboratory observations with graded sands showed that suspended sand sizes are smaller than the original sand bed, increasing with the height above the bed. This is due to discrimination with respect to entrainment, although once in suspension finer and coarser grain fractions show similar concentration profiles (Nielsen, 1983, 1992).

Dibajnia & Watanabe (1996) observed that the transport rate of the finer fraction is reduced by the presence of coarse sediments, which behave as if there were no finer sand present, thus indicating that hiding effects takes place; for sand mixtures sheet flow occurs

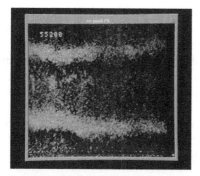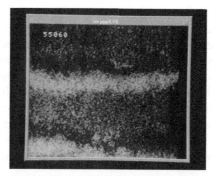

Figure 4.5 Top view of a rippled sandy bed characterized by a bimodal mixture at two different instants within the wave cycle.

under lower oscillatory velocities than for uniform sand; a higher velocity is required to wash out ripples formed with widely-graded sand than in the case of a uniform one.

Bimodal sediment mixtures in flume experiments usually consist of sand sized sediment ($D_{50} < 2$ mm) and gravel sized sediment ($D_{50} > 2$ mm) in different proportions. Bimodal mixtures are used for modelling the effects of sediment sorting processes such as downstream fining (e.g. Paola *et al.*, 1992) and the gravel-sand transition (e.g. Sambrook-Smith & Ferguson, 1996) and to estimate transport rates where local size information is insufficient to resolve the transport rate on a fraction by fraction basis (e.g. Wilcock, 1998).

Sambrook-Smith & Ferguson (1996) identified that Froude scaling of mixed gravel/sand sized sediment can be problematic since finer gravel-sized sediments may need to be replaced by sand which may have different shape characteristics leading to different packing arrangements, and sand sized sediment may need to be replaced by silt-sized sediments which will exhibit cohesive characteristics. Therefore, experiments may need to retain the original prototype sediment size distribution and use lower depths and higher slopes to achieve equivalent shear stresses. Although this may introduce some hydraulic distortion it is possible to maintain grain Reynolds numbers $Re^* = (u^* D_{90})/v_k$ indicative of fully rough turbulent flow and subcritical flow conditions. Warburton & Davies (1998) suggest that if the grain Reynolds number is greater than about 70 then rough turbulent flow will occur which is independent of viscous effects; following this principle it is suggested that to fulfil approximate dynamic similarity the following requirements must be satisfied:

* Sediment density is the same in the model and prototype;
* Model grain sizes are scaled geometrically;
* Water is used as the fluid;
* Bed slope is the same in the model and prototype;
* Flow is rough-turbulent; and
* Froude number similarity.

It may prove difficult to achieve all these requirements in practice, e.g. scaling all the grain-sizes in a widely-graded mixture at a small geometric scale might result in the finest grains exhibiting cohesive properties. Sometimes the bed slope is increased in small flumes to compensate for a shallow water depth in order for the Shields parameter to be maintained. One problem with this is that the near-bed turbulence changes in very shallow flows, so that threshold condition also changes.

Experiments using mixtures of sand and gravel have shown that the transport rates of sand and gravel depend on the proportion of each size fraction present in the bed which controls the availability of each size fraction (Wilcock, 1998). However, the critical shear stress required to transport the gravel-sized fraction also decreases as the sand content of the mixture increases (Wilcock, 1998). Moreover, the structure of the bed changes from clast-supported to matrix-supported when the sand fills up the pores of the gravel entirely, affecting the bed roughness and bedforms.

Cohesive sediments and mud

Sediment mixtures with a fraction of clay particles larger than about 10% have cohesive properties because electro-static forces comparable to or higher than the gravity forces are acting between the particles. Consequently, the sediment particles do not behave as individual particles but tend to stick together forming aggregates known as flocs, whose size and settling velocity are much larger than those of the individual particles. Settling velocities of individual particles (deflocculated) are in the range of 0.1 to 1 mm/s.

The grain size scale of the American Geophysical Union for sediments with particle sizes smaller than 2 mm consists of about 13 subclasses ranging from very coarse sand to very fine clay. Herein, five somewhat broader subclasses are distinguished:

- Coarse sand (non-cohesive)   0.5 to 2 mm   (500 to 2000 μm)
- Fine sand (non-cohesive)   0.062 to 0.5 mm   (62 to 500 μm)
- Coarse silt (sometimes cohesive)   0.032 to 0.062 mm   (32 to 62 μm)
- Fine silt (weakly cohesive)   0.008 to 0.032 mm   (8 to 32 μm)
- Clay+very fine silt (very cohesive)   <0.008 mm   (<8 μm)

The cohesive fraction with clay and very fine silt is taken to consist of particles with diameters smaller than 8 μm (clay-silt dominated fraction). Bed samples consisting of mixtures of clay, silt and sand can be classified as in Table 4.1.

The transport of bed material particles may be in the form of either bed-load or bed-load plus suspended load, depending on the size of the bed material particles and

Table 4.1 Types of sand-mud mixtures.

Type of sediment	Organic material	Clay + fine silt (<8 μm)	Silt (8 to 62 μm)	Sand (>62 μm)
Sand (non-cohesive)	0%	0%	0%	100%
Muddy Sand (weakly-cohesive)	0–10%	0–5%	20–40%	60–80%
Sandy Mud (cohesive)	0–10%	5–10%	30–60%	60–30%
Mud (cohesive)	0–20%	10–20%	50–70%	0–10%
Silty Mud (cohesive)	0–20%	10–40%	60–80%	0%
Clayey Mud (cohesive)	0–20%	40–60%	40–60%	0%

the flow conditions. The suspended load may also include some wash load (usually, clay-silt dominated fraction smaller than 8 µm). Cohesive properties become effective when the clay-silt dominated fraction is larger than about 10%. The critical mud fraction (particles <62 µm) will be a factor 2 to 3 larger (20% to 30%).

The use of physical models for studies involving cohesive sediments is not current practice in most hydraulic laboratories, but their use has been promoted both in France (principally by Sogreah) and in China. These experiences are described below.

Physical modelling with mud in France

The use of mud in physical models was developed in France by Dr. Claude Migniot between 1955 and 1987 along two main lines: (a) a proper characterization of mechanical properties of natural muds in various environments, and (b) for prediction of siltation in harbour basins, navigation channels and water intakes located mainly in estuaries (current–dominated environment).

Standard laboratory tests on natural muds were developed by Migniot in the fifties and early sixties to characterize their mechanical properties. A large database has been gathered in fluvial, coastal, estuarial and industrial environments and reported (in French) in two major publications in La Houille Blanche (Migniot, 1968, 1989). These results have been disseminated internationally by Migniot & Hamm (1990) and Hamm & Migniot (1994) and also through two European projects (MAST-G6M and MAST2-G8M projects) between 1990 and 1995 (MAST-G6M, 1993). Such tests included:

- Deposition tests in calm water with deflocculated (in fresh water) and flocculated (in salty water) sediments;
- Consolidation tests including changes in the rheological properties of the deposit;
- Erosion and deposition tests under current action on consolidated beds;
- Mobility of fluid mud in calm water.

The technical details of the test procedures and instrumentation have been documented during the G6-M Coastal Morphodynamics European project (MAST G6M, 1993). Experiments of flume measurement of mud transport on a flat bottom under steady and alternating currents was fully documented in the MAST-G8M project (Viguier et al., 1994a).

Two important points should be stressed here. First, natural mud samples are used and care should be taken with organic matters which induce rapid evolution of mud characteristics when transferred in the lab. Usually, organic matters are neutralized before the experiments. This is a strong limitation when they play a major role in the field. Secondly, adequate and specialized instrumentation should be used; most particularly to measure vertical concentration profiles and rheological properties in deposits.

Physical scale model techniques have been developed empirically by Migniot for solving practical siltation problems in a series of about ten port developments. Such models were highly distorted by a factor 4 to 12. Two kinds of sediments have been used in such models depending on the percentage of clays. When this percentage was low, bakelite powder was used. On the other hand when it was significant, treated natural mud was used. This treatment was aiming at getting a correct settling velocity and critical shear stress for erosion in similitude (Caillat, 1987). It included an elimination of the coarse part of the sediment and a chemical treatment of the mud and the water with sodium pyrophosphate to reduce the cohesion of the mud.

Recent models of this kind included: the Mont-Saint-Michel sedimentation study where silts and fine sands were dominant (Viguier et al., 2002); and the Loire estuary studies where three models were built and used to predict siltation rates in port developments (Caillat, 1987). One of these models (siltation in a transverse trench) was documented during the MAST-G8M project in order to be used as a benchmark for numerical siltation models (Viguier et al., 1994b). Sogreah has also used a physical scale model to study the behaviour of ships progressing in fluid mud (Brossard et al., 1990).

However, such mud models have not been run at Sogreah for the last five years for two reasons. Firstly, mechanical properties of muds are measured in-situ and secondly 3D numerical morphodynamic models of mud transport have now reached a similar or even better level of efficiency to predict siltation at a much lower cost and with higher flexibility. The case where such a scale model could be envisaged in the near future is to reproduce channel meandering on silty tidal flats where numerical models are still not able to make reliable long-term predictions. In such cases we recommend the use of very light artificial sediments like sawdust (see Mont Saint-Michel study).

Physical modelling of fine sediments in China

Most of the Chinese water systems are rich of fine sediments (high to hyper concentrations) and sedimentation of fines is a serious problem at many sites. These problems have been studied by scale models based on the theoretical analyses of Guoren Dou (1998). According to his approach, all particles and all forms of sediment transport can be reproduced simultaneously in one model. The main difficulty is that the time scale of bed deformation related to suspended load is very different from that related to bed load. His analysis indicated that the difference in time scales of bed deformation was caused by the fact that the similarity of incipient motion was not considered for suspended load transport models, while the similarity of particle settling was not considered for bed load transport models. If both similarity criteria are satisfied, the difference in time scales of bed deformation will disappear.

Guoren Dou (1998) present a set of scaling laws including the Froude scale, the settling velocity scale, the sediment transport scale and the time scale involved.

The settling velocity scale is: $N_{ws} = N_U (N_l/N_h)^{-1} = (N_h)^{0.5}(N_l/N_h)^{-1}$.

Here N_l, N_h, are model geometric scale factors in the horizontal and vertical directions (distorted model), and N_{ws} and N_U are scale factors for settling velocity and current speed.

Depth and length scales are of the order of 100 and hence the settling velocity scale is of the order of 10. Given the fact that the suspended sediments in the prototype are already very fine, the model sediment would have to be very fine, resulting in the use of light-weight materials. Often, polystyrene with a base density of 1060 kg/m³ or bakelite powder with a base density of 1450 kg/m³ have been used as model sediment. For example, to study the sedimentation problems related to the fluctuating back water zone of the Three-Gorges Dam in the Yangtze River, a scale model ($N_l = 250$, $N_b = 100$) with a length of 800 m was made. Scouring and deposition patterns in the scale models have been found to show very good agreement with prototype patterns.

4.4.2 Scaling laws

General principles

The basic philosophy for movable-bed models can be formulated as ensuring that the relative magnitudes of all dominant processes are the same in model and prototype. Preferably, the scale model should be validated using field (prototype) data, but often this is not feasible and large-scale model results are used as prototype data. The scaling must be considered for both the hydrodynamics and the sediment dynamics, and correct scaling of the former does not necessarily lead to correct scaling of the latter.

Two "tricks of the trade" are sometimes used to assist with obtaining scale-similarity:

- Use of a distorted-scale model, in which the vertical scale-factor is smaller than the horizontal scale-factor (i.e. vertical exaggeration);
- Use of low-density model sediment, which gives an extra variable that can be utilised to obtain scale-similarity of more than one parameter.

Despite their apparent advantages, both of these techniques introduce extra uncertainty into the interpretation of model results at prototype scale, so they should be avoided if possible, or used with a full understanding of the consequences if they are adopted. Further discussion of these techniques is given later in this section. However, we will start by considering an undistorted model, and natural-density sediment.

The model scale factor, N_X, of a physical parameter X is defined as the ratio of the prototype value of X to the model value of X. Thus the geometric model scale factor in an undistorted model is N_l, where l is any characteristic length (e.g. defining the bathymetry, or the size of a structure). In a distorted-scale model N_l is the geometric scale factor in the horizontal direction, and N_b is the (usually smaller) geometric model scale factor in the vertical direction (including water depth h). In undistorted models, $N_b = N_l$, while in distorted models the vertical exaggeration is N_l / N_b. Similarly, model scale factors can be defined for any other physical variables; for example, N_t, N_{ws}, N_{qb} are the scale factors for time t, settling velocity w_s, and bedload transport rate q_b.

Hydrodynamics: Froude scaling

The most widely used, and generally applicable, scaling law used for the hydrodynamics of free-surface flows in physical models is Froude scaling. If the geometric scale (i.e. the ratio of lengths l in the prototype to those in the model) of an undistorted model is N_l, then with Froude scaling all times are scaled by N_t, where $N_t = (N_l)^{1/2}$. This scaling law was developed in the 19th century by William Froude on the basis of his pioneering model tests of ship dynamics in towing tanks. Furthermore, it can be shown from consideration of the momentum equations that Froude scaling is applicable in general to free-surface flows where the dominant controlling force is gravity. This scaling follows from the need to maintain constant ratios between the various terms in the equations of motion in order to have dynamic similarity between model and prototype. The relationship $N_t = (N_l)^{1/2}$ is required because the gravitational acceleration, g (units $[LT^{-2}]$), is normally the same in model and prototype (with the exception of centrifuges for modelling of soil mechanics). This scaling is equivalent to the requirement that the Froude number $U/(gh)^{1/2}$ is the same at model and prototype scales, where U is (current or wave-orbital) velocity, and h is water depth. For open channel flow the Froude number is a very important parameter describing the character of the flow: tranquil or shooting flow, whether disturbances can propagate upstream or not and the magnitude of variations in the surface level when the flow is disturbed. For surface gravity waves no similar important dimensionless Froude number can be defined – even though surface gravity waves are some of the hydrodynamic phenomena best reproduced in a scale model based on Froude's model laws.

Many hydrodynamic quantities are correctly scaled by Froude scaling:

- Current speeds scale as $N_U = (N_l)^{1/2}$
- Wave heights and wavelengths scale as N_l
- Wave periods scale as $N_T = (N_l)^{1/2}$
- Wave orbital velocities scale as $N_U = (N_l)^{1/2}$
- Wave orbital excursion amplitudes scale as N_l
- Keulegan-Carpenter numbers applicable to sediments or rock-protection ($KC = U_w T/D_g$) and structures ($KC = U_w T/D$) are identical in model and prototype provided that the bed material is geometrically scaled, because they depend only on the ratio of bed material (or structure) size to wave orbital excursion. Here D_g is grain or rock diameter, D_s is structure diameter, U_w and T are wave orbital velocity amplitude and period.
- Current drag coefficients in rough turbulent flow are identical in model and prototype provided that the bed material is geometrically scaled, because they depend only on the ratio of bed roughness to water depth.
- Wave friction factors in rough turbulent flow are identical in model and prototype provided that bed material is geometrically scaled, because they depend only on the ratio of bed roughness to wave orbital excursion.
- Bed shear-stresses τ in rough turbulent flow scale as velocity-squared provided that the bed material is geometrically scaled and water densities are identical, and hence $N_\tau = N_U^2 = N_l$.

However, in a scale model only a single force can be correctly reproduced at a time. For example it is not possible to reproduce the forces of gravity and viscosity in the same model when water is the fluid in the model as well as in the prototype. In a scaled-down physical model a Froude model law gives reduced flow velocities compared to the prototype, while a correct scaling of the viscous forces (maintaining the same value of the Reynolds number in model and prototype) would require an increase of the flow velocities in the model compared to the prototype. The Reynolds number is consequently reduced in the Froude model by a factor of $N_l^{3/2}$, and the viscous effects will be more pronounced in the model than in the prototype. This makes it more likely that the model flow is smooth turbulent or transitional ($Re^* < 70$), in which case the last four bullet points above would not hold.

In general wave phenomena like shoaling, refraction and diffraction follow the Froude model law, and the wave field over a given bathymetry can be quite accurately reproduced in a model. The onset of wave breaking, the breaker type and the wave decay due to breaking can be reproduced, but depends on the magnitude of the model waves. Even though small waves may be reproduced satisfactory outside the surf zone, the viscous and surface tension effects can be of significance for the characteristics of wave and flow phenomena inside the surf zone. This must be addressed specifically when planning a model test involving breaking waves. The distribution of the flow field over a bathymetry, for example the distribution of the specific discharge (discharge per metre width) and the mean flow velocity over a river cross section will normally be satisfactory reproduced, still provided that the Reynolds number is sufficiently high to ensure turbulent flow in model.

Furthermore, some sediment dynamic quantities are correctly scaled by Froude scaling, provided that the bed material is geometrically scaled and has the same density (relative to the water) as the prototype:

* Shields parameters of sediments and rock-protection in rough turbulent flow are identical in model and prototype, because they depend on the ratio of bed shear-stress to grain size, if the density-ratio s and gravity g are the same in model and prototype;
* The threshold Shields parameter of rock-protection is (almost) identical in model and prototype provided that the model rock is sufficiently large that viscous effects can be ignored. However, this only holds true (to within 10%) for non-dimensional grainsize $D^* > 120$, which for quartz in fresh water at 20°C corresponds to $D > 5$ mm. For smaller model rocks, and for sandy sediments, the threshold Shields parameter varies with D^* and will generally not correspond between model and prototype.

Quantities which depend on more than one dimensional physical variable are not easy to scale down from prototype to model, as they do not scale correctly with Froude scaling. These include cases in which a significant role is played (at model and/or prototype scale) by any of the following processes:

* Viscous forces. In cases where the viscous forces are dominant at prototype scale, the Reynolds number rather than the Froude number should be made equal in model and prototype. Examples include:
  o laminar wave boundary layers (weak waves)
  o permeability effects on percolation through sediment beds in beach dynamics and offshore foundations
  o settling velocities of fine sediments in suspension. These are very dependent on the Reynolds number, but the settling is due to the action of gravity and it can therefore not be reproduced accurately in a scale model unless the sediment density is changed.
  o wave friction factors in smooth and transitional turbulent flow
  o current drag coefficients in smooth and transitional turbulent flow

- Hot water is sometimes used to reduce the kinematic viscosity of water at model scale (but has a limited applicability). Air is sometimes used as a substitute for water in models at very small scale.
- Surface tension forces. If waves are modelled at a very small scale, capillary waves (dominated by surface tension) might be of similar wavelength to gravity waves, which will disturb the wave dynamics. The level of the water table within a beach depends on surface tension, and percolation in the swash zone on a shingle beach will normally not be reproduced correctly. Droplet formation depends on surface tension, although this is not usually important for sediment dynamics.
- Coriolis accelerations. The effect of the Earth's rotation is important only at prototype scales larger than a few kilometres and time scales larger than a few hours, e.g. in the tidal dynamics of coastal seas, large estuaries and inlets. These effects can be modelled using rotating (Coriolis) turntables, although it is unusual for these to be used for sediment dynamics.
- Electro-chemical forces between sediment grains. If medium sand in the prototype is attempted to be reproduced in, say a 1:50 scale model, the grain size in the model will be of the order 10 micron. In this range the inter-granular forces begin to be of significance, for example as a stabilising force for a resting grain. In addition to the effects of viscosity this imposes an effective limit to the grain size of prototype material which can be reproduced directly in a scaled down model to produce quantitative results.

Scaling of sediment transport

The scaling of sediment transport rate does not usually follow directly from Froude scaling. In general, the scaling of sediment transport requires knowledge (or more usually an assumption) of the form of a sediment transport formula that is appropriate to the scenario being modelled. If the sediment transport formula is written in terms of a product of powers of the input parameters, then the scaling law can be derived as the same product of the scale-factors for the input parameters. Multiplicative coefficients drop out of the scale relationship, so it is not necessary to know these. However, some sediment transport formulae cannot be written in terms of power laws, in which case the ratio of the full expression at prototype and model scales must be taken. Again, leading multiplicative coefficients drop out, but some internal coefficients do not, in which case a sensitivity analysis to the coefficients is necessary to give a range of results at prototype scale. In addition, a sensitivity analysis to the use of other possible sediment transport formulae should be made.

An exceptional case in which Froude scaling can be applied directly is that of bedload transport of shingle (gravel or cobbles) in which the model shingle is larger than 5 mm and has the same density ratio s as the prototype. In this case, the Shields parameters of incident currents and waves are identical in model and prototype, and so is the threshold Shields parameter. Since the non-dimensional bedload transport is often expressed as a function of only the incident and threshold Shields parameters, the Froude scaling holds good. The sediment transport at a point has (volumetric) units of [m²/s], and hence the scale factor for the bedload transport rate q_b is $N_{qb} = N_l^2 \cdot N_t^{-1} = N_l^{3/2}$. The long-shore bedload sediment transport rate integrated across the beach profile Q_b has (volumetric) units of [m³/s], and hence the scale factor for the bedload transport rate is $N_{Qb} = N_l^3 \cdot N_t^{-1} = N_l^{5/2}$. In both cases the results can be converted to mass units by multiplying by the sediment density. The shape and angularity of the prototype shingle should also be reproduced at model scale as far as possible.

However, it is often the case that the choice of model shingle is limited by the available sizes of material, so that the prototype size cannot be scaled exactly geometrically. In addition, there may be differences in the densities of the model and prototype shingle, and/or the densities of the water (e.g. sea water in prototype, fresh water in model). In these cases, some adjustments of the scaling must be made. This will normally rely on making use of the dependencies found in theoretical expressions for the bedload transport. This introduces an element of uncertainty into the scaling, because different formulae have different dependencies, especially for long-shore transport. The full scaling laws for such cases are discussed by Hughes (1993).

Scaling of morphological evolution

Provided that the sediment transport rates and directions are accurately reproduced in the model, the modelled morphological evolution should represent the prototype situation. This is because the morphological evolution is governed by the divergence of the sediment transport rate, which applies equally at model and prototype scale. For *suspended* transport it is important that lag effects are reproduced correctly in the model. This can be achieved by scaling such that $Z = w_s/u^*$ is the same in model and prototype – however, it is not in general possible to scale by the Froude number and by Z simultaneously, so a compromise scaling is often necessary. An alternative is to use low-density sediment to achieve this, but the caveats noted in section 4.4.3 should be observed.

The rate of evolution of the bed is governed by the morphological time-scale T_M. For the "ideal case" of Froude scaling for *bedload* transport defined above, the morphological timescale scales as $N_{TM} = (N_l)^{1/2}$. If this ideal scaling cannot be obtained, the time scale for morphological evolution will deviate from the Froude scaling law. In some cases, for example scour tests, the geometry of the morphological evolution will be similar in the model and the prototype, and an estimate of the time scale may be made from the sediment transport rate estimated for the two cases. However, there are severe limitations to this technique, for example if the ratios between bed and suspended load are not the same or if the geometries of the morphological evolution are not similar.

Scaling river sediment models

Modelling of flow in rivers requires Froude scaling: $N_{Uf} = (N_h)^{0.5}$ with $Uf =$ depth-mean flow velocity, $h =$ water depth. Correct flow resistance based on Chézy law requires: $N_U = N_C (N_h)^{0.5} (N_{Sl})^{0.5}$ with $C =$ Chézy value and $Sl =$ slope. Distorted models can be included by using different bed slopes. Provided the modelled flow is hydraulically rough-turbulent, and Manning's Law applies, correct modelling of the Chézy value in hydraulic rough flow requires $N_C = (N_h)^{1/6}(N_{ks})^{-1/6}$ with $k_s =$ Nikuradse roughness.

Scale modelling of morphological processes in the case of *dominant bed load* requires that the grain Shields parameter $\theta = \mu h Sl/[(s-1)D_{50}]$ with $\mu =$ ripple factor, $s =$ relative density of sediment and $D_{50} =$ median sediment size is reproduced.

The Shields parameter should be the same in model and prototype: $N_\theta = 1$, yielding: $N_\mu N_{Sl} N_b = N_{s-1} N_{D50}$. Using the Chézy law, the ideal velocity scale can be derived: $(N_{Uf})^2 = (N_{s-1}) (N_{D50}) (N_C)^2 (N_\mu)^{-1}$. Assuming $N_{s-1} = 1$ for sand and $N_\mu = 1$ (geometrical bed forms in model and prototype), it follows that the ideal velocity scale is given by: $N_{Uf} = (N_{D50})^{0.5} N_C$.

Assuming that the dimensionless sand transport parameter is given by: $q_b/[g^{0.5} D_{50}^{1.5} (s-1)^{0.5}]$, the sand transport follows from: $N_{qb} = (N_{D50})^{1.5} (N_{s-1})^{0.5}$ with $q_b = $ bed-load sand transport rate in m²/s.

It should be noted that one cannot *directly* control the geometry of the bed forms in the model to achieve a scale of $N_\mu = 1$. One can only select suitable model sediment subject to all the other constraints. The result is that frequently compromises have to be made. In some cases model sediment is used that differs in density from the pro-totype sediment. The selection of appropriate light-weight model sediment depends upon the materials available and the scales being used (section 4.4.3)

Generally, the ideal velocity scale for morphology will differ from the Froude velocity scale, resulting in scale effects. Deviation from Froude scaling is acceptable for conditions with small Froude numbers. Scale modelling of *dominant suspended sediment transport* requires that the scale factor for the suspension number $Z = w_s/u*$ (with $w_s = $ settling velocity and $u* = $ friction velocity) is equal to unity. Thus $N_{ws} = N_{uf*}$ or $N_{ws} = N_{Uf} (N_C)^{-1}$ or $N_{Uf} = N_{ws} N_C$.

Scale errors can to some extent be corrected for by using distorted scale models. However, if the physical model involves natural river banks then this may constrain the selection of the distortion of the horizontal and vertical spatial model scales. Equi-librium channels of different sizes have different aspect ratios and this implies a dis-tortion between the horizontal and vertical scales.

If the purpose of the model is to predict a new equilibrium morphology following engineering or other works then an issue is the rate of change of morphology in the model relative to the prototype. A morphological timescale can be defined to express this: morphological time scale $N_{Tm} = N_{vol}/N_{Qs}$, where N_{vol} is the volumetric scale for the sediment and N_{Qs} is the scale for Q_s, the sediment transport rate integrated across the cross-section.

The requirement for the model to reproduce morphological change in the prototype in a reasonable period of time may require larger sediment concentrations in the model in the prototype. Detailed information on river scale models is given by Janssen *et al.* (1979). An example of a river sediment model is given in Section 4.10.1.

Scaling coastal sediment models

For coastal scale models it is often important to attain similarity of the cross-shore equi-librium bed profiles between prototype and model, particularly in the surf zone (Hughes & Fowler, 1990). This means that the dimensionless parameters describing equilibrium profile behaviour should be the same in model and prototype. The most dominant mode of transport being either bed load or suspended load (depending on wave condi-tions and bed material) should be represented correctly. Both undistorted models and distorted models have been used in scale modelling. Ensuring similarity in an undis-torted model is less complicated than in a distorted model. Dean (1985) reasoned that, in an undistorted model with a sand bed, the fall trajectory of a suspended particle must be geometrically similar to the equivalent prototype trajectory and fall with a time proportional to the prototype fall time. This can be accomplished by ensuring similarity of the fall velocity parameter between the prototype and the model, which is only feasible in an undistorted model. Distorted models can be used when the reproduction of the bulk erosion volume (dune erosion volume; Vellinga, 1986) is most important, while the precise reproduction of the equilibrium profile in the entire surf zone is less important. The vertical exaggeration of the distorted model is N_l/N_h.

Correct representation of the bed forms in nature and model requires that the Shields parameter $f(U_w)^2/[(s-1)gD_{50}]$ is equal in both cases (f = friction factor, U_w = wave orbital velocity). Thus:

$$N_\theta = 1 \text{ or } (N_{Uf})^2 = N_{s-1} N_{D50} (N_f)^{-1} \tag{4.1}$$

Based on linear wave theory: $U_w = \pi HT^{-1} \sinh(k_w h)$ with $k_w = 2\pi/L$. In shallow water: $\sinh(k_w h) = k_w h$. Thus:

$$U_w = 2\pi^2 HhT^{-1} L^{-1} \tag{4.2}$$

Using $N_H = N_L = N_h$ and the scale requirement for correct wave dynamics and breaking, $N_T = (N_h)^{0.5}(N_l/N_h)$, it follows that:

$$N_U = (N_h)^{0.5}(N_l/N_h)^{-1} \tag{4.3}$$

Combining Equations (4.1) and (4.3), yields: $(N_{Uf})^2 = N_h (N_l/N_h)^2 N_{D50} (N_f)^{-1}$ or,

$$N_{D50} = N_h N_f (N_{s-1})^{-1} (N_l/N_h)^{-2} \tag{4.4}$$

Using sand in a non-distorted model, if follows that:

$$N_{D50} = N_h N_f \tag{4.5}$$

The friction factor scale generally is much smaller than 1 ($N_f \ll 1$) for storm conditions (flat bed in prototype and ripples in model; $f_p \cong 0.01$ and $f_m \cong 0.05$ or $N_f \cong 0.2$). This behaviour is favourable for modelling purposes as it leads to a less-fine model sediment. For example: $N_h = 25$, $N_f = 0.2$ and thus $N_{D50} = 5$. However, it should be appreciated that ripple-related processes occurring in the model might not occur in the prototype. Furthermore, regular waves should not be used in a scale model because the ripples may induce offshore-directed sand transport against the wave direction (in the case of non-breaking waves).

As regards fully developed sand transport, it essential that (ideally) the dimensionless excess shear stress or (more simply, but less correctly) the excess velocity ($U_w - U_{cr}$) is reproduced correctly (U_{cr} = critical orbital velocity for initiation of motion). This requires that $f(U_w - U_{cr})^2/[(s-1)gD_{50}]$ is equal in both model and prototype or:

$$(N_{uf-ucr})^2 = N_{s-1} N_{D50} (N_f)^{-1} \tag{4.6}$$

Scale effects will be relatively small as long as $Uf \gg U_{cr}$ in the scale model and relatively large for $Uf \cong U_{cr}$ or $Uf < U_{cr}$ in the scale model. In the latter case no sediment motion will occur in the model. This may occur when relatively coarse sand is used in the scale model. These scale errors generally are largest outside the surf zone (offshore slope of outer bar). Scale conditions close to initiation of motion should be avoided as much as possible.

The Shields parameter is most important in shoaling waves with dominant sand transport close to the bed. This parameter is less important in the surf zone with strongly breaking waves.

Scaling laws for beach and dune erosion by storm waves

A number of scaling laws have been proposed for beach and dune erosion, based on different assumptions about the relationships between the variables. Some of these are for undistorted models ($N_l = N_h$) and some for distorted models ($N_l/N_h > 1$). These are summarised here with: N_{ws} = settling velocity scale, N_{TM} = morphological time scale, N_A = erosion area scale.

Noda (1972) has performed various *beach erosion* experiments in scale models (distorted and undistorted) with regular waves focussing on relatively coarse sand (>0.5 mm; initial slope of 1 to 25). Based on analysis of quasi-equilibrium bed profiles, he proposed the following scaling laws:

$$N_{D50} = (N_h)^{0.55} \tag{4.7}$$

$$N_l/N_h = (N_h)^{0.32} \tag{4.8}$$

Ito & Tsuchiya (1984, 1986, 1988) and Ito *et al.*, (1995) have done detailed studies of beach erosion profiles in quasi-equilibrium conditions under low and high waves. The scaling laws derived from these undistorted scale model (regular waves) series are:

$$N_{D50} = (N_h)^{0.83} \text{ for } N_h < 2.2 \tag{4.9}$$

$$N_{D50} = 1.7(N_h)^{0.2} \text{ for } N_h \geq 2.2 \tag{4.10}$$

$$N_{TM} = (N_h)^{0.5} \tag{4.11}$$

Based on analysis of scale model results for *dune erosion*, the set of scaling laws proposed by Van Rijn *et al.*, (2010) is:

$$(N_l/N_h) = (N_h)^{0.28} (N_{D50})^{-0.5} (N_{s-1})^{-0.5} \tag{4.12}$$

$$N_A = N_l N_h \tag{4.13}$$

$$N_{TM} = (N_h)^{0.56} = (N_l/N_h)^2 (N_{D50})^1 (N_{s-1})^1 \tag{4.14}$$

According to Vellinga (1986), the scaling laws for dune erosion read as:

$$(N_l/N_h) = (N_h)^{0.28} (N_{ws})^{-0.56} \tag{4.15}$$

$$N_A = N_l N_h \tag{4.16}$$

$$N_{TM} = (N_h)^{0.5} = (N_l/N_h) (N_h)^{0.22} (N_{ws})^{0.56} \tag{4.17}$$

A comparison of these different scaling laws for field and lab data-sets is presented in Section 4.10.3.

Scaling of scour pit development

The scaling laws governing scour pit development are as follows. It is not usually possible to scale the sediment size geometrically, because the behaviour of the sediment would change radically. For example, sand of diameter 0.2 mm (200 microns) scaled geometrically at 1:50 would have a model grain diameter of only 4 microns, corresponding to the borderline between very fine silt and coarse clay. The behaviour of the model sediment would be dominated by cohesive electro-chemical forces, and hence would be unrepresentative of the prototype sand. However, it is known that, provided that the ratio of grain size to pile diameter is smaller than 0.02, the dimensions of the scour pit (depth, extent, etc) scale geometrically with the pile diameter. In addition, the current speeds must be chosen such that U_c/U_{cr} is the same in the model as in the full-scale conditions. For waves, the criterion is that the Keulegan-Carpenter number $KC = U_w T/D$ is the same in the model as in the full-scale conditions. In many cases sand can itself be used as the "model" sediment. In a few cases it may be helpful to use larger-grained but lower-density materials such as coal or plastics, to reduce the value of U_{cr} to a lower value, but the caveats noted in section should be observed.

See Section 4.10.4 for an example of modelling scour and scour protection around an offshore wind-turbine monopile.

Scaling of rock scour protection

Different scaling laws apply to rock scour protection. In this case, all the length dimensions *can* be scaled geometrically (at ratio 1: N_l), and the times and velocities must be scaled at ratio 1: $N_l^{1/2}$, following Froude scaling. Other quantities that scale correctly with Froude scaling (for model rock of equal density with the prototype) are:

- The KC number of the pile ($KC = U_w T/D$);
- The KC number of the rocks ($KC = U_w T/D_g$);
- The current drag coefficient (provided that the model rocks are large enough for the current flow to be hydraulically rough-turbulent);
- The wavelength of water waves;
- The wave friction factor and drag coefficient (provided that the model rocks are large enough for the wave flow to be hydraulically rough-turbulent);
- The wave added-mass coefficient (provided that the model Reynolds number is large enough);
- The Shields parameter of the rocks, provided that the model rocks are larger than about 5 mm. If the model rock and water densities are different to prototype, corrections to the Shields parameter can be made.

Hence model results for rock protection scale all the important governing non-dimensional groups correctly, and can be interpreted with confidence at prototype scale. In cases where the model rock has a different density from the prototype rock, some of the quantities above will not scale exactly and corrections must be applied. The ratio $H_s/[(s-1)D_g]$ is often used for scaling purposes for rock, especially for surface-piercing rock structures.

One quantity that does not scale correctly is the Reynolds number, because the viscosity of the water is the same in model and prototype. However, providing that the scale chosen is sufficiently large, the viscous forces will be small (compared to other forces) and the lack of Reynolds scaling can be safely ignored. Similarly, permeability of the seabed is not scaled correctly, and hence model tests of soil-mechanics properties need to be scaled differently. The permeability of the rock protection can nonetheless be scaled approximately when using a Froude-scaled model.

Furthermore, the protection of the underlying sand may be influenced by the flow in the porous protection layer, which might not be modelled correctly. For example, Darcy's (viscous flow) law might apply in the model but not in the prototype, where turbulent flow might occur in the pore-spaces between rocks. See section 4.10.4 for an example.

Choosing the best geometrical model scale

Large-scale models (i.e. small geometrical scale factor) have advantages over small-scale models (large scale factor) in terms of the certainty of scaling the results up to full-scale, due to smaller scale-factors being used and the smaller scaling errors due to extraneous processes, especially due to their relatively small unscaled effects of viscosity.

These benefits must be set against the additional expense, labour and time involved in needing larger quantities of water and sediment. Furthermore, the reduced accessibility (more difficult to get close to features of interest) longer setting-up times (water filling and draining; filling, moulding and removing sediment) and longer run times (resulting from the smaller geometrical, and hence time, scale factor) must be considered.

The relative merits must therefore be considered of having more reliable results, but for a small number of test cases of limited (prototype) duration, versus a larger number of test cases and longer (prototype) test duration but with greater uncertainty in scaling-up the results obtained.

Choosing the model sediment

The following factors should be considered when choosing the model sediment:

*   If the scaled-down sediment has a diameter d_{50} less than about 60–100 μm, electrochemical cohesive forces become important and would spuriously influence the sediment dynamics;
*   The presence of even very small quantities of clay (5–10%) could cause cohesion of larger grains at model scale, even though they would not at prototype scale. This can also apply to biological cohesion;
*   The scaled-down sediment should be in the same hydrodynamic roughness regime as the prototype – in many cases a sand bed that would be hydrodynamically rough at prototype scale will become hydrodynamically smooth at model scale (spurious effect of viscosity);

- Low density sediments can be considered if they (a) help to preserve scale similarity, (b) allow sediment to move under modelled flows, or (c) provide a larger permeability in cases where this is important;
- However, low density sediments may also have disadvantages such as (a) having different shaped grains, (b) floating on the water surface due to surface tension, and (c) making the scaling laws less transparent;
- The source of model sediment. Choose between (a) natural beach/river sediment, which might reproduce the range of grain-sizes realistically, but might also contain mud, shell fragments, biological debris, etc, and (b) industrial sands (used in metal casting) which have very well-controlled grain-size distributions, can be obtained with various colourings, and can be delivered in small or large quantities, but are expensive and come in a limited number of sizes.

4.4.3   Low-density (model) sediments

The adoption of low-density sediments (otherwise known as light-weight sediments) is sometimes required in experiments at reduced scales. Indeed, if velocity in the prototype is low, prototype sand used in the model may not meet incipient motion conditions easily.

A model utilising low-density sediment allows both the Reynolds number and the densimetric Froude number to maintain the same value in the prototype and in the model (see Hughes, 1993 for a detailed review of sediment transport model criteria). However, there are also disadvantages, because the geometrical length, fall speed and relative density scales are not satisfied and consequently undesired scale effects are introduced. For instance, the omission of the relative density criterion and the similarity of the sediment density to that of the water could lead to considerable magnification of sediment suspension: this means that in this case suspended transport will be overestimated at the expense of bedload transport. In other cases, the sediment characteristics might be chosen specifically to satisfy realistic modelling of the suspended sediment transport rate, but then bedload transport and bedform characteristics might not be simulated correctly.

Because the geometric length scale is not respected and the sediments are larger than they should be, the particle displacement induced by the flow is relatively less than what would be expected. Bedform characteristics are therefore underestimated, with obvious consequences also on roughness predictions. However, laboratory models including localized effects can still be carried out making use of low-density sediments, as, for example, in the case of scour processes. Ideally, in order to judge the reasonableness of the result of model tests when low-density sediments are used, the test results of low-density sediments should be compared with that of prototype sand.

Use of low-density sediment in river models

In general, the flow in the prototype is rough turbulent and, as a consequence, the corresponding flow in the model should also be rough turbulent. This can be expressed in terms of the Reynolds number. If we assume initially that the hydraulic roughness is determined by the sediment size then the condition on the flow being rough turbulent becomes a condition on the grain size Reynolds number. To satisfy this condition one may need to use sediment in the model that is larger but has a lower density than that in the prototype. In practice this occurs when the scale factor N_{Dg} for the grain diameter D_{50} satisfies the condition

$$N_{D50} > (70/Re^{\cdot})^{\frac{2}{3}}$$

(4.18)

where

$$Re^{\cdot} = \frac{u^{\cdot} D_{50}}{v_k}$$

(4.19)

Widely available low-density sediments are characterised by a relative density not much different from unity, spherically shaped, made of high density plastic or resins or synthetic composite materials, such as Bakelite, PVC, polystyrene, nylon, etc. In principle, any material whose density is smaller than the prototype, which commonly has a specific gravity of 2.65, and greater than that of water can be used. In practice, cost and practical problems normally limit the selection of material.

Sediment movement is commonly divided into bed load and suspended load. It may be necessary to ensure that sediment in the model moves in the same mode as in the prototype. This is often characterised by the suspension number $Z = w_s/u^*$ where u^* is the shear velocity and w_s is the settling velocity. In general, u^* in the model will be smaller than that in the prototype and as a consequence it may be necessary to resort to a low-density sediment to provide the appropriate value of Z.

The use of low-density sediments means that the sediment particles in the model are larger than that implied by the linear scale of the model. The implication is that the skin roughness of the sediment bed will be relatively greater in the model than in the prototype. This has to be allowed for in the selection of the model scales. In general, the angle of repose of low-density sediments will be different from most naturally occurring sediments. Care should therefore be taken in the use of low-density sediments in situations where the angle of repose is important in determining the geometry of the bed.

In the derivation above it was assumed that the hydraulic roughness is dominated by the roughness of the sediment grains. If significant bed features are present then this may affect the validity of the model. In general, problems arise when the bed features in the model differ in either character or size from those in the prototype.

It is not necessary to use only one type of model sediment. Where a range of sediment sizes are present in the prototype it is possible to use a low-density sediment to represent the finer fractions while a natural sand can be used to represent the coarser fractions.

Use of low-density sediments in wave models

To reproduce mobile beds under waves one needs to ensure similarity of the Shields parameter:

Table 4.2 Low-density materials within specified ranges of specific gravity.

Material	Usual range of specific gravity
Acrylonite Butadienne Styrene (ABS)	1.22
Bakelite	1.3 to 1.45
Coal	1.37 to 1.61
Light weight aggregate	1.7
Perspex	1.18 to 1.19
Polyamidic resins (nylon)	1.16
Polystyrene	1.035 to 1.05
PVC	1.14 to 1.25
Sawdust treated with asphalt	1.05
Sand of Loire	1.5
Silica (Ground)	Not reported
Walnut shells (Ground)	1.33
Wood (Granulated Obeche)	1.10

$$\theta = \frac{\rho(u^*)^2}{g(\rho_s - \rho)d} \tag{4.20}$$

As u^* will, in general, be smaller in the model than in the prototype it may be necessary to use a low-density sediment with a reduced value of ρ_s. However, the effect of inertia forces under waves is more important for low-density grains than for natural grains of stone or ore density. This can put an effective limit to how low a density that can be applied in a model test.

The issues related to bed forms discussed for river models above are equally an issue for the use of low-density sediments for wave models.

Types of low-density sediments

A wide range of materials have been used in physical models. Bettess (1990) lists 13 of them in Table 4.2.

Properties, experiences, advantages, problems

In practice the use of low density is often difficult. Due to their low density, there is a tendency for the sediment to float. This may be a problem during the initial filling of a model with sediment. It can also be a problem with river models where sediment has to be introduced at the upstream boundary during a test.

Some low-density materials, such as chopped wood, deteriorate with time. When using coal it has been observed that there is a tendency for the material to become segregated in the model depending upon the shape of the particles. Although suggested by some textbooks, the use of activated carbon or other type of organic material should be avoided. Indeed such kinds of low-density sediments change their physical characteristics after some time, so that their mobility properties are also affected.

Problems have been reported by some practitioners in wave models using low-density sediments where there is wave breaking. This probably reflects a general issue of representing sediment movement in a region of breaking waves. It is to be noted that none of the standard methods for scaling wave models take into account the conditions associated with breaking waves.

In certain circumstances, (e.g. to achieve mobility or a realistic permeability) low-density sediments confer distinct advantages. However, they should only be used with a full appreciation of the theoretical and practical problems associated with them.

4.4.4   Boundary conditions on sediment

In tests where sediment transport is expected to be significant, the mobile sediment bed may lower significantly during the test, unless sediment is fed into the system at the upstream end. Most simply, this can be achieved with a sediment feeder system, in which sediment is introduced continuously at the upstream end, but not re-circulated. Simple feeder systems will not require manual intervention. Differences between input and output cause a change of bed slope, after which it disappears. For long-running experiments it is advantageous to re-circulate the sediment. There are two types of sediment recirculation systems (Sutherland *et al.*, 1999b): online recirculation systems, where sediment captured at the downstream end is fed back continuously to the upstream end (Sutherland *et al.*, 1999a) and offline recirculation systems, where sediment is collected at the downstream end of the flume and a similar volume is input at the upstream end (Kamphuis & Kooistra, 1990; Badiei *et al.*, 1994). An online system operates all the time, and recirculates what comes out of the downdrift end, so the method can be used for long tests and does not depend on modelling to determine the volume of sediment to be transported. However, an online system does not allow the transport rate to be measured accurately, except by periodic sampling or by a sediment flux measurement in the recirculation pipe. Moreover, some water will be recirculated with the sediment, which will adjust the discharge of the flume. An offline system can be easy to operate and can have a good temporal resolution. However, offline systems require more manual intervention, may not input the same volume of sediment that is being output and may require short test durations, if the sediment traps fill up quickly.

Wilcock (2001) discusses the importance of the mechanism used for sediment transport in gravel bed experiments for feeding sediment into the flume. The two common approaches are to add sediment using a constant rate and composition or to recirculate entrained sediment and use this as the feed supply. Parker & Wilcock (1993) demonstrated that the upstream sediment supply conditions affect the final equilibrium state. Where sediment is added, the flume adjusts the surface composition, flow depth and slope to transport the imposed load, which isolates the effect of transport rate on bed composition and slope, but does not represent the variation in transport composition with flow strength. In contrast, when using sediment recirculation the sediment feed depends on the flow conditions and the relative mobility of different size fractions in the bed material and allows variation in both transport rate and grain size with flow strength, but the final bed surface and transport rate depend on the initial conditions of the bed surface. Sediment can be thoroughly mixed and screeded flat, but this does not necessarily replicate a natural bed surface found in the field.

Results from sediment feed experiments (e.g. Kuhnle, 1989; Dietrich *et al.*, 1989; Wilcock & McArdell, 1993) shows that at low transport rates the surface grain size distribution becomes coarser since the less mobile grains accumulate on the bed surface. At higher transport rates the differences in grain mobility are reduced and the bed surface does not become coarser. Experiments using recirculating sediment (Wilcock & McArdell, 1997) show that as flow increases, larger grain sizes become more mobile and this increase in grain size of the transported material also leads to a coarsening of the bed surface due to a kinematic sorting process (Parker & Klingeman, 1982, Wilcock & Southard, 1989). For dunes, Kleinhans (2005) demonstrates that the upstream supply condition (feed or recirculation) strongly affects the vertical sorting by the dunes. Feed flumes must eventually transport all the fed sediment, so that the depth-averaged composition of the dunes plus the layer reworked by infrequent deep troughs becomes equal to that of the feed sediment. In recirculation flumes on the other hand the coarse sediment, which usually is slightly less mobile than the fine initially, is worked down below the dune troughs to form a strong armouring layer while the transported sediment fines considerably. Therefore it is important to consider whether the experimental configuration is appropriate to the problem being investigated.

4.5 MEASURING AND OBSERVATION TECHNIQUES

4.5.1 Measurement techniques for sediment transport

A wide range of devices may be used in a physical model test of sediment dynamics. It is important to understand the method of operation and the limitations of the instruments that are to be used before commencing a set of experiments. This is beyond the scope of the present chapter, which merely describes typical uses. An overall introduction to the measurement of sediment transport is given by van Rijn (2007). This report gives an overview of the types of instruments (particularly the most recent types) used to measure sediments, particularly bedload sediment transport, suspended sediment and bathymetry or morphology.

4.5.2 Instrumentation for bedload transport

Luminescent tracers

The use of luminescent tracers in the flume investigation of sandy-gravel bed-load movement has been developed at VITUKI. The method can be recommended for the visual observation of incipient motion of selected grain-size fractions in mixed-grain bed materials. Traditionally quasi-homogeneous grain-size fractions have been used as bed materials in laboratory flumes to determine their critical flow velocities or shear stresses. However, in most natural stream channels, well-sorted bed materials are to be found in which the observation of incipient grain motion according to their diameters are rather difficult if not impossible.

The application of the luminescent tracing method (Rákóczi, 1987) distinguishes 3–4 different grain-size fractions in a mixture by marking them with special dyes of various colours shining brilliantly under ultraviolet light. The procedure of fixing the luminescent dyes on the surface of the previously washed and dried grains is similar to that applied in field sediment tracing studies. For the laboratory flume experiments, however, the bed materials are mixed according to the prototype or prescribed compositions and their marked fractions are represented in the necessary percentages. The walls of the glass flume are temporarily covered in order to keep visible light outside as much as possible and the flume is illuminated by UV-lamps from above. The grains of the marked fractions can easily be seen and distinguished both from each other and from the unmarked grains on the flume bed surface. Consequently, the incipient and advanced phases of their displacement can be observed visually and/or documented by photographic or video means.

The main advantages of applying luminescent tracer techniques in laboratory sediment studies are that the critical flow velocities determined in this way are more realistic than the results of the traditional single-fraction studies, since the former represent the effect of the grain-size composition on the threshold of sediment movement. Secondly, a good insight can be obtained into the process of selective erosion of bed materials, almost always present in nature and being an inevitable component of armour layer development.

Bedload sampler

Bedload samplers use a mechanical trap to intercept sediment particles that are in transport close to the bed. Each has a limited width and height, which may mean that it collects a portion of the suspended sediment transport as well as the bedload. Popular bedload instrument include the Helly-Smith sampler for sand, the Karolyi for gravel, and the Delft Nile bed load sampler. Further details are given by van Rijn (2007) and www.coastalwiki.org/coastalwiki/.

Conductivity Concentration Meter

The CCM system is an instrument for the concentration measurements of sand-water mixtures with varying conductivity of liquid. The principle is based on the conductivity change of a sand-water mixture due to the variation of the quantity of non-conductive sand present in the measured area. Used to measure suspended sand transport in the sheet-flow layer in wave conditions in large wave flume.

With a second (slave) unit the varying water-conductivity can be continuously compensated. The instrument can be modified for 600 Hz response time. The probes consist of a 4 electrode system composed of 0.5×1 mm Ø platinum. The probe head is shown in Figure 4.6.

Concentration through the attenuation of sound

The Ultra High Concentration Meter (UHCM) determines particle concentration by measuring the attenuation of sound between a transmitter and receiver placed 11 mm apart. The instrument is specifically applicable in flows with high sediment concentrations of up to 1200 g/l for China clay and up to 400 g/l for 0.2 mm sand. Each sensor has a diameter of 9 mm and width of 6 mm. The output signal of the probe is linearly proportional to the concentration. A schematic diagram and photograph are shown in Figure 4.7.

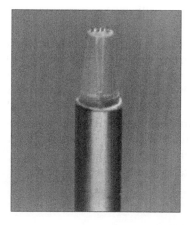

Figure 4.6 Conductivity Concentration Meter (courtesy University of Twente).

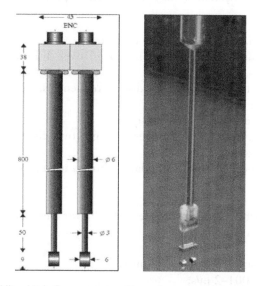

Figure 4.7 Ultra High Concentration Meter (courtesy University of Twente).

4.5.3 Instrumentation for suspended transport

Acoustic backscatter system, coherent acoustic doppler velocity profiler and ripple profiler

Principle of operation

The concept of using acoustic diagnostics in the underwater environment is attractive and straightforward. A pulse of high frequency sound, typically in the range 0.5–5 MHz in frequency, and millimetric/centimetric in length, is transmitted from a directional sound, source usually mounted within about a metre above the bed, and the backscattered signal gated into range bins and digitised. As the pulse propagates down towards the bed, sediment in suspension backscatters a proportion of the sound and the bed generally returns a strong echo. The backscattered signal amplitude and rate of change of phase respectively provide profiles of suspended sediment particle size and concentration, and the three orthogonal components of flow, while the bed echo provides the time history of the bed location. The objective of using acoustics has been to obtain profile measurements of the suspended sediment and flow with sufficient spatial and temporal resolution to allow turbulence and intra-wave processes to be probed, which coupled with the bedform morphology observations, provide sedimentologists and coastal engineers with new measuring capabilities to advance our understanding of sediment entrainment and transport. Further details are given in a review article by Thorne & Hanes (2002) and more recent applications are given in Thorne *et al.,* (2009) and Hurther & Lemmin (2008).

Typical working characteristics

Acoustic Backscatter System (ABS) for vertical profiles of particle size and concentration:

- Sediment grain size: 40 µm to 4 mm diameter;
- Sediment concentrations: 0.01 g/l to 20 g/l over 1 m, or more over shorter range;
- Transducers: 10 to 20 mm ceramic discs. Frequencies, from 500 kHz to 5 MHz
- Range: 1 to 4 m;
- Transmission rate: 100 Hz max pulse rate;
- Data hardware averaging: Ensembles averaged over time by powers of 2 (up to 64) before storage;
- Range cells: 256 cells, 10 mm standard, options usually 2.5 mm, 5 mm, 20 mm and 40 mm.

Coherent Acoustic Doppler Profile (ADVP) for vertical profile of the three orthogonal components of flow:

- Sediment grain size: 10–500 µm diameter;
- Velocity range 0.001–2 m/s;
- Transducers: 10 to 20 mm ceramic discs, rectangular. Frequencies, from 1 MHz to 3 MHz;
- Range: 0.1 to 1 m;
- Transmission rate: up to several KHz;
- Data hardware averaging: Ensembles averaged over time by powers of 2 (up to 256) before storage;
- Range cells: up to 256 cells, 3 mm standard, options usually 3–40 mm.

Ripple Profiler (RP) for transects of the bed profile obtained by scanning a single beam over an arc in a vertical plane and sampling regularly:

- Transducers: 20–50 mm ceramic discs. Frequencies, from 500 kHz to 2 MHz;
- Range: 2–5 m;

- Transmission rate: 100 Hz;
- Range cells: 100 cells, mm standard;
- Bed detection algorithm, usually mm vertical resolution.

Optical Backscatter System

An Optical Backscatter Sensor (OBS) measures turbidity and suspended solids concentrations by detecting infra-red light scattered from suspended matter (van Rijn, 2007). The response of an OBS sensor strongly depends on the size, composition and shape of the suspended particles. Battisto *et al.,* (1999) show that the OBS response to clay of 0.002 mm is 50 times greater than to sand of 0.1 mm of the same concentration. Hence, each sensor has to be calibrated using sediment from the site of interest. The measurement range for sand particles (in water free of silt and mud) is about 1 to 100 kg/m^3. The sampling frequency is usually 2 Hz.

Particle Image Velocimetry / Particle Tracking Velocimetry

Particle Image Velocimetry (PIV) and Particle Tracking Velocimetry (PTV) are optical methods that involve using a camera to capture more than one image of particles in the water. The velocity of the particles is determined by the distance moved and time between photographs. The particles can be mobile sediment particles or tracer particles (often small and/or low density) that should follow the fluid motion. Recently, efforts have been made to measure the flow speed of the fluid (represented by the movement of micron-sized seeding particles) and of sediment in suspension from the same image. PIV measures the average velocity of the particles within a small area, while PTV tracks individual particles between images. The principle of PIV is outlined below:

- Small (usually micron-sized) seeding particles are introduced into the flow;
- A pulsed laser light sheet illuminates a planar field of interest within the particle-laden flow at predetermined time intervals;
- A digital camera records images of the particles within the illuminated plane of interest;
- Pairs of particle images are analysed to reveal the distance travelled by the particles in the time separating the recordings;
- Velocities are determined by dividing the distance travelled by the time interval between pulses;
- Seeding particles and sediment particles are separated using size and intensity of image. This only works if there is a clear difference in the sizes of the seeding and sediment particles;
- The seeding particles are assumed to follow the flow, so a 2D map of two components of fluid and suspended sediment velocity are obtained; and
- Suspended sediment concentration may also be estimated.

The range of velocities that can be measured depends on the timing between images and the displacement of images within the image. These parameters need to be optimised for each experiment.

Bottle and trap samplers

Bottle and trap samplers collect a water sample from the experiment to determine the local sediment concentration. The fluid velocity at the intake would ideally be the same as that of the surrounding fluid. Some samplers simply fill a bottle (van Rijn 2007) while others use a pump system and may have a number of bottles for sampling at different times. For example, the Aberdeen Oscillatory Flow Tunnel has four multi-bottle pump sampling systems that rotate in phase with the regular oscillatory flows generated. This allows sediment concentration to be determined at different locations (such as different elevations above the bed) and at regular phases through a flow cycle. Details of the sampler can be found in O'Donoghue & Wright (2004).

4.5.4   Instrumentation for measuring bathymetry

Measuring in the wet or the dry

Measurements of bathymetry can be made in the dry, by lowering the water level before measurement is made and raising it after, or in the wet by measuring the seabed through the water column. Measuring in the dry requires additional time for the draining and filling of the facility and the act of draining can wash out some of the fine structure of the morphology, for example by smoothing off the crest of ripples. Draining a facility is sometimes done overnight, so that a particularly slow rate of draining can be used, thereby disturbing the seabed features as little as possible and giving time for any small residual pools of water in the bottom of bedforms to drain away. If an experiment is stopped in order to measure bathymetry the suspended sediment will settle and this may have a slight effect on the sediment transport, although this is not a problem in the case of gravel studies. Examples of methods that can only be used in the dry include:

- Terrestrial laser scanners;
- Some forms of acoustic and optical sensors that can be included in a bed profiling system can scan transects of the bed profile or can measure continuously at a point;
- Touch-sensitive bed profiler which is good for capturing the profile of rocks (whether in a rubble mound breakwater or a cobble beach) where a dragged wheel may stick or smooth out the profile too much;
- The movement of armour stones or units can be determined by taking photographs of a section of the armour both before and after a test, from precisely the same position. A comparison of the photographs reveals the stones or units that have been moved.

Where draining a big flume is too long a job, profiles must be taken with it full. The traditional approach to doing this is to deploy a bed level measuring device on a carriage that moves along a set of rails and scans the elevation in a straight line. An additional requirement is that the bed level measuring device can operate from the dry part of the beach to the wet part.

An optical sensor may require the water to be clear and it may take some time for suspended sediment to settle out. In a small to medium size facility it may be quicker to drain and fill a facility than to wait for the water to clear sufficiently.

Repeatability of the experiments is really difficult, because the packing of the sand is very important for the ripple formation at small scale. In large flumes there is not usually time to run the same experiment twice. Ripples can give a spurious measurement in an experiment, when an average level is desired. Traditionally ripples could be smoothed physically, but more recently the ripples can be filtered out in post processing.

Bed profilers

The most common form of beach profile sensor for large wave flumes uses a wheel (or rollers) mounted on a long arm that is dragged along the profile on a carriage, as shown in Figure 4.8. The carriage is driven along the flume and the cross-shore position, x, and the angle of the arm, α, are logged. It is a simple process to convert trolley position and arm angle into bed position in the horizontal and vertical. Such profilers can measure underwater and above the water surface in a single profile and are relatively insensitive to changes in water level, water temperature or pollution in the water. They can profile over rock armour as well as sand, providing that the wheel is large enough to run over the abrupt changes in elevation that will be encountered (in which case some of the detail of the rock armour will be smoothed out). They are also relatively quick, for example the measuring system in the GWK can profile up to 250 m of the flume at a rate of 10 m/minute. However, the dead weight of the wheel and arm will cause the profiler to penetrate a short distance into the sand. For example, the measuring system in the GWK penetrates between 2 and 20 mm into a sand bed.

Other touch-sensitive bed level measuring devices that can be used for measuring across the interface between beach (or structure) and the seabed include a pressure-sensitive foot that is stepped across a profile (as used at HR Wallingford and shown in Figure 4.9). If there is no need to measure above the water level, or in very shallow water, a range of other probe heads can be used with a stepping or traversing profiler,

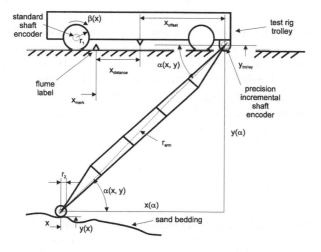

Figure 4.8 Geometry of dragged wheel bed profiler.

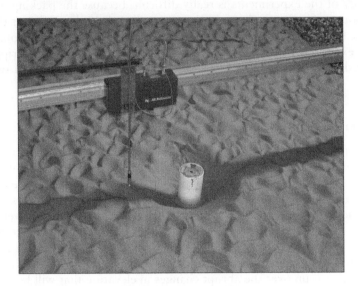

Figure 4.9 Touch-sensitive bed profiler surveying radial profile of scour around a monopile (the top part of which has been removed) (courtesy of HR Wallingford).

including capacitance, acoustic (ultrasonic) and laser probes. Such probes can also be installed in a fixed position to give a time series of bed level at a position.

Laser scanners

Commercial laser scanners can be used to collect point-clouds of (x, y, z) bathymetric data. Typically thousands of points can be sampled in each second, with an accuracy that depends on the setting, range and number of times that each point is sampled. One commercially-available system can sample 5,000 points per second with an rms accuracy of 1.4 mm at a range of 10 m, based on sampling each point 4 times. Increasing the number of times each point is sampled reduces the rms error, but increases the length of time it takes to scan an area. For example, this system can be used to survey an area of 10 m by 5 m at an average spatial resolution of 10 mm by 10 mm in 1000 seconds, sampling each point 10 times for a rms error of 0.9 mm.

The points sampled in the scanner's coordinate system can be transformed into the coordinates of the physical model by scanning in targets (normally spheres) that are in known locations (determined by traditional survey techniques). A 3D elevation model produced using a laser scanner is shown in Figure 4.10.

Digital photogrammetry

A Digital Elevation Model (DEM) can be created via digital photogrammetry – the use of two simultaneous overhead photographs of a physical model. Brasington & Smart (2003) demonstrated the use of digital photogrammetry to derive high-resolution DEMs of an eroding floodplain every 15 minutes through a 240 minute simulation.

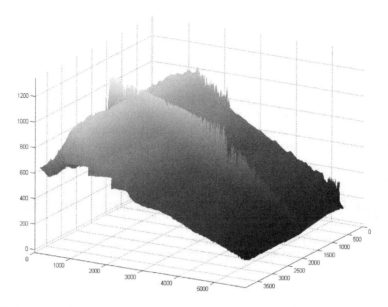

Figure 4.10 Physical model bathymetry measured using a laser scanner (courtesy of HR Wallingford).

A vertical precision of 1.2 mm was obtained, which produced a threshold level of change detection of ±3 mm at 95% confidence level.

Optical techniques

To overcome the limits of many of the traditional instruments used in the presence of an erodible bottom, which are often quite invasive and require that the experiments is stopped, so that the evolution of the phenomenon cannot be investigated at a small time scale, optical techniques can be used. Two different measurement techniques, 2D and 3D, have been developed at University of Catania for the experimental investigation of small scale morphodynamics (Faraci *et al.*, 2000; Baglio *et al.*, 2001). In their original version such techniques have been applied to the investigation of the wave-induced scour at the base of a cylindrical pile, but their use has been extended also to the measurement of ripple formation and evolution (Faraci & Foti, 2002).

The above measurement systems are made up by: (i) a laser, (ii) a cylindrical lens for 2D, and a diffractive lens for 3D; (iii) and one or two cameras, for the 2D and 3D system, respectively. In the 2D case, a 'sheet' of laser light is projected onto a target object that produces a reflected light that follows the contours of the object. By a proper calibration of the images, a simple correspondence between the horizontal and the vertical dimension of the measuring object and the image size can be obtained. In the 3D case, a grid of light points is projected onto the target object and using a purposely developed model, which allows removing the classical hypotheses of stereoscopic vision theory, the 3D dimensions of the object can be obtained. It has been demonstrated that, at least at a scale of O (30 cm^2), the accuracy of such techniques is comparable to that of micrometers.

Among the advantages of the aforementioned techniques, the possibility to project a measuring point onto the even sandy bottom, by means of the structured light approach, and that of performing continuous time evolution analyses, in a completely non-invasive way, must be mentioned. One of the main difficulties with such techniques is related to the fact that the free surface generates spurious reflection, which must be avoided. Therefore, the observation of the phenomenon and the projection of the laser light are usually made through glass windows in side walls of the tank. However, recently the possibility to use a waterproof system, i.e. to bring both the light source and the image acquisition system below water level, has been investigated and some preliminary results have been obtained (Foti *et al.*, 2007).

Echo-sounders

A commercial 3D echo sounder could be used in water depth greater than about 0.5 m. This could be used to produce 3D underwater bathymetries. The range measurements needed for a laboratory test, even in a large facility, mean that the instrument would have to be carefully chosen and set up to operate at the shortest possible distance.

Marked pins

A traditional technique for measuring scour depths is to install graduated pins in the seabed. The pins are photographed and the depth of scour determined from the visible length of pin exposed by the scour. This can be done without draining the facility.

4.6   PROCEDURES FOR PERFORMANCE OF TESTS

4.6.1   Calibration

Calibration is an essential preliminary to the main test series. It should be carried out on individual instruments (current metres, depth gauges, wave gauges, sediment transport instrumentation, etc.) and on the flow and/or wave generation (settings to produce a given set of currents or wave heights and periods).

Calibration of current metres is usually performed in a towing tank, where the speed of the carriage forms the primary standard. For depth gauges and wave gauges, calibration is usually performed by altering the relative heights of the gauge and a still water surface through a range of positions covering all values likely to be encountered plus a safety margin. Calibration of electronic instruments for measuring suspended sediment concentration (e.g. ABS) should be made against a series of known concentrations of the actual sediment used (preferably with the same mix of grain sizes that are found in suspension, as opposed to in the bed). The accuracy of the calibration and variability on repeat calibrations should be noted.

4.6.2   Planning a test series

A provisional table of planned tests should be drawn up at the outset. However, flexibility is required, as it might be necessary to alter it as the test series proceeds, due to tests taking longer than anticipated, or in response to results from the first few tests.

Usually in sediment-related modelling there are many variables that could be altered (water depth, current speed, wave height/period/direction, sediment size/grading, initial bathymetries, etc.) It is often impractical to cover all combinations of a series of values of each variable. For example, all combinations of three each of depth, current speed, wave height, wave period and sediment grainsize would yield 243 tests. Instead, a subset is usually chosen, for example, a baseline set of the most representative value of each variable is chosen, and one variable at a time is varied. In the example given above, this would yield 11 tests – a much more manageable number.

4.6.3   Costs and times

The cost and time needed to complete the programme of work needs to be assessed early on in the project, in conjunction with the planning of the test programme. Often this will have been done at the proposal stage, but it is worth re-visiting once the project is started. Costs include staff-time and materials. The cost of sand in a large facility can be considerable, and time for delivery should be allowed. Artificial bed materials can be even more expensive, especially if they need grading or grinding to the desired grain-size distribution.

Commissioning costs include communication time with the client, design of model and test series, construction costs including materials, time for calibrationsand time for preliminary tests. Costs per test include staff time to turn around a test (set-up for each test and duration of test).

Analysis, interpretation and reporting time should not be under-estimated. This is often one-third of the total cost and time. De-commissioning costs should also be considered, for removing the materials, instruments, etc in readiness for the next project.

Total cost for tests = commissioning + analysis/interpretation + reporting + de-commissioning + cost per test.

Total elapsed time is calculated similarly.

4.6.4   Construction

The following notes provide details of methods used by HYDRALAB participants to prepare a sand bed prior to conducting an experiment.

Cleaning of sediment

The sand used in big installations is cleaned just once at the beginning by reproducing long wave time series and by the initial cleaning when deploying the sand inside the flume full of water. The sand or gravel in a smaller facility may be cleaned by hand. This process does not distort the D_{50} of the sediment due the fact that the removed part is mainly dust and really fine sediment.

Installing sand in a facility

Dry sand generates problems when filling a flume or basin with sediment. If a big amount of dry sand is deployed in a flume and later on it is filled with water there will be a really important amount of air within the grains. These air bubbles will distort all the measurements that are intended to be done. The best way to deploy dry sand in a flume is with it full of water, when the sand is deployed inside the full tank there will be no air bubbles within the grains. When all the sand has been deposited into the flume and has settled through the water column, the flume can be drained and the sand bed shaped to the desired profile.

In cases when it is not practical to deposit sand through the water column, some water should still be left in the floor of the facility and a hose may be used to partially wet the sediment as it is deposited. When the sand had been deposited and shaped to the desired profile the entire sand bed should be soaked for ideally two days by over-filling the facility with water. Some waves or flow should then be run to help the sand bed settle before the start of the full test series.

Sand may be deposited in a large facility by being dumped directly from a lorry or digger. In smaller facilities the sand is usually hand distributed by the help of sacks or buckets and spread with a rake or shovel.

Levelling of sand to a desired profile

The desired initial sand bed profile is often achieved in a large flume by levelling the sand following lines previously painted on the walls of the flume. Alternatively, relatively thin wooden templates may be installed against the flume walls at the correct elevations, although this will cause local distortions to the bathymetry, which restricts the useful width of the flume. The sand bed may be screeded if it is sufficiently small for a beam to stretch across it, overlapping the hard bottom by about 0.5 m. This generally limits the width of screeded beds to 4 m or 5 m at most. Screeding involves moving a smooth beam backwards and forwards over a saturated sand bed until the surface of the bed is smooth. An example of a screeded sand bed is shown in Figure 4.11.

Transitions to a sand bed

Transitions between hard bottom and a sediment bed often cause a local scouring effect caused by a change in bed shear stress between the two surfaces. This problem can often be minimised by gluing grains of sand or gravel or other roughness elements, such as a wire mesh, to the hard bottom. In big flumes there is usually a flat part with sediment before the sloping sediment beach that helps to avoid the transition problem. The flat part of the profile work as transition area, allowing the waves to become stable. Changes in depth or roughness have natural transitions in big installations by the use of initial slopes and a length area in which the waves evolve before meeting the study area. When the study area is going to be affected also by

Figure 4.11 Screeded sand bed around 0.14 m diameter monopile base (courtesy of HR Wallingford).

currents, the transition length should be 10 to 20 times the water depth before having the study area in order to avoid boundary problems. When running long experiments in sand pits there is often a loss of sediment.

4.7 RESULTS

Modern data handling and storage technology can cope with large volumes of data, ensuring that all raw data can be kept and subsequently re-processed if necessary. In order to do this the data format must be well documented so that data can be retrieved even when the original hardware and/or software used to collect the data has become obsolete. Increasingly it also means that meta-data (information about the data) is stored electronically with the raw data. Alternatively the meta-data on the experiment may be recorded in a report, which should be stored with the data in a commonly readable format (such as pdf). At least one backup copy of all data should be kept in a separate building from the master copy.

Different laboratories have their own systems for acquiring data. The development of enabling technologies (Wells *et al.*, 2009) offers the potential to use the experience of the wider scientific community by adopting common standards and developing them for the needs of hydraulic laboratories.

Documentation must be available at different levels:

* The infrastructure;
* An experimental project (for instance a Transnational Access project);
* An experiment, and
* A record from a particular device in an experiment.

The infrastructure should be described, both in terms of the physical facility (description, photographs, dimensions, purpose, track record) but also the organisation that hosts it, with an emphasis on chain of responsibility for the facility and its databases. Available instrumentation should be described as should the standard coordinate system.

Details of an experiment may include funding organisations, key staff (particularly principle investigators and visiting researchers), purpose of the experiments, dates and duration of experimental programme, budget (likely to be confidential) and a listing of all publications including laboratory notebooks, data reports, the database of results and subsequent publications (if known about) such as conference proceedings, journal papers and theses.

Details of an experiment will include the location and type of instruments, parameters measured (with their units), sampling rates and duration, directory and file names of raw data and the target conditions run. This level will also contain free remarks on the experiment (e.g. quality appreciation, observations on particular events, calibration procedures).

Records from a particular device in an experiment should include:

* The time of all measurements in the record relative to the time origin of the experiment;
* The absolute time origin, with date-time (e.g. 2006–05–30 14:02:12);

- Names of the recorded physical quantities;
- Calibration parameters needed to translate the recorded data into the physical quantities;
- Complementary information needed to control and reproduce the experimental procedure (such as sensor name, gains, motor speed, options depending on the instrument used).

Storing information from all these different levels together and in a systematic way should allow a dataset to be used by scientists or engineers who were not involved in the data collection or storage.

Data analysis should be transparent, with key equations used set out in detail with references given to the source of the equations to justify their use. Established procedures should be followed where possible. Filtering procedures should also be documented, so that the analysed records can be derived independently by other researchers.

Quality control is important in ensuring that experiments are trusted and re-used. It is based on thorough documentation of the facility, the measuring devices and the procedures used in running experiments, storing and analysing the data. An experienced auditor should be able to trace data from their source to their calibrated form with confidence, based on the records left by the experimenter. Laboratories should ideally operate a procedure for the control of documents and records (which includes experimental data).

Interpretation of the results from the physical modelling of sediment dynamics should be conducted by, or under the supervision of, a senior researcher or consultant. Any interpretation should be illustrated using examples from the dataset and should refer to the limitations of the model, as well as its strengths.

4.8   REPORTING

A typical report on physical modelling of sediment dynamics for a consultancy application will include (as appropriate):

- Project description and problem definition (statement of objectives, Terms of Reference or Scope of Work);
- Project methodology (including appreciation of physical processes modelled and neglected with justification, and reasoned choice of scaling method and geometric scale selected);
- Description of test facility, model set-up, method of construction and instrumentation;
- Test programme;
- Description of test results (including sources and sizes of errors);
- Conclusions and recommendations.

Figures and tables will typically include (as appropriate):

- Table of test parameters;
- Tables of test results;
- Grain-size distribution curve of sediment;

- Fall-velocity distribution curve of sediment;
- Drawing of test facility, with dimensions;
- Plots of bed morphology;
- Photographic documentation.

Research reports will also typically contain many of the above items. The original log-book and data sets should be archived for future reference.

4.9   UNCERTAINTY ASSESSMENT

It is important that all the possible sources of error in undertaking physical model tests are identified, and (where possible) quantified. The main sources of error (assuming that the most important processes have been correctly identified at the outset) are: scaling errors, laboratory effects, and instrument accuracy. Of these, the latter is usually the smallest. If the individual errors can be quantified, they can be combined in the usual manner by summing variances, assuming that they are statistically independent:

Total error = [(scaling error)^2 + (lab error)^2 + (instrument accuracy)^2]$^{1/2}$

4.9.1   Scaling errors

For experiments performed at small scale, there will usually be some processes which have not been scaled correctly. For example, a model based on Froude scaling will not be correctly scaled for viscous effects. Drag coefficients for currents, and friction factors for waves, contain both a bluff-body drag and viscous drag terms, and the relative magnitude of the viscous terms can be calculated from the theoretical expressions for them. These can then be interpreted in terms of scale errors for the velocities (for given forcing) and the bed shear-stresses. The scale error for, say, sediment transport rate can be assessed from these errors, if the relationship is known (roughly) from theory or the model tests themselves. Effects of surface tension or sediment cohesion can be assessed by similar methods if appropriate.

Errors may also be introduced if the model sediment grain size is larger than the geometrically-scaled grain size. This will enhance the roughness of the bed. However, provided that the scaling laws are fully understood, this can be allowed for in scaling results up to prototype. Large-scale facilities (large wave flumes, oscillating water tunnels) suffer less from scale errors than small-scale facilities, although laboratory errors may still be present. In particular, the Reynolds numbers may be sufficiently large that spurious effects of viscosity are negligible.

4.9.2   Laboratory effects

Even if the scale errors can be minimised there are still errors arising from performing experiments in laboratory facilities, when compared to the case in the field. Examples include:

- Limited breadth/depth ratios in current flumes;
- Blockage effects when testing flow around structures;

- Use of regular instead of irregular waves (but not often done now);
- Omission of long-shore processes in wave flumes;
- Omission of vertical velocities and longitudinal variation in OWTs;
- Insufficiently fully-developed waves in flumes and basins, giving unrealistic surface elevations and kinematics;
- Inadequate absorption of reflected waves in flumes and basins;
- Transverse oscillations in wave flumes;
- Over-steep beaches in wave flumes and basins;
- Inadequate re-circulation of long-shore currents in wave basins (e.g. allowing them to return around the offshore part of the model, instead of pumping and re-introducing);
- Inadequate re-circulation of sediments, leading to deficits in the bed;
- Incomplete grain-size distribution of sediments;
- Unnatural compaction of sediments in the bed.

Distorted models may be necessary to overcome practical limitations, but will often introduce laboratory (or scale) errors.

4.9.3   Instrument accuracy

The accuracy of the instrument will usually be quoted in the manufacturer's specification. Accuracy should not be confused with resolution – a read-out displayed to 4 significant figures is not necessarily accurate to 0.01%. The prudent experimenter will make his own calibrations against a primary standard at the time of the experiments, which will remove some of the error. Variations in repeat calibration coefficients can give a measure of the instrument errors. In the absence of such information, a rough guide is that instruments will give results accurate to about 1%, and will log or display the results to 0.1%.

4.10   EXAMPLES

4.10.1   Scour downstream of river barrages

The depth and extent of scour downstream of river barrages can be quite significant, especially in sand-bed rivers, even if the damming height is low. Designers often require the assistance of movable-bed physical models for checking the planned type and dimensions of bed protection, especially the necessary downstream length of it.

The first task of river barrage modellers is find the hydraulically optimum site for the structure within the selected river reach, the laboratory investigation of a planned river barrage usually consists of two main parts:

- Modelling of a several kilometres long river stretch with fixed banks and movable bed and
- Modelling of a several metres wide stream-wise section of the barrage with movable steel gate structure and concrete foundation built into a straight glass flume with movable bed material.

The *river model* is constructed using moderate vertical distortions, e.g. in 1:200 horizontal and 1:50 vertical scales. If space permits, the model can be built undistorted, e.g. in 1:75 scale. The latter solution is preferred, especially if the barrage complex contains also a navigation lock and a run-of-river hydropower plant. During the tests, the gates of the barrage, the lock and the turbines are operated in various combinations and in different stages of opening and/or closing them. Detailed flow-velocity measurements are carried out at a number of operating variants under various flow conditions. Thus, useful conclusions can be drawn concerning the stability of the prototype grains on the riverbed and for the calculation of the expected scour depths.

As for the grain sizes and densities of the model bed material, in most cases they can not be converted using the model scales. Natural sand may be used in the model if the prototype river has a gravel bed. In case of finer beds, granular materials lighter than sand might be tried in the model, possibly a mixture of them, however, usually repeated preliminary tests have to be conducted. Anyway, river models, serving the general layout, give only approximate indication of the scour dimensions downstream of the barrage.

In this respect, more detailed information can be expected from the *glass flume tests* of undistorted barrage models e.g. with 1:75 scale in a 1 m wide channel. Here almost any gate operations and flow conditions can be reproduced and the grain-size composition of the prototype bed material can sufficiently be approximated and controlled. The phases of the scouring process can be well documented by photo and/or video techniques and the shape and dimensions of the obtained scour holes measured.

4.10.2 Cross-shore beach profiles

Beach profile development under storm surge

Recommendations for the planning, performance and analysis of 2DV large-scale experiments in wave flumes are given. Experiments in large wave flumes such as in the Großer Wellenkanal (GWK) in Hannover, Germany, and the Delta Flume in De Voorst, Netherlands, have the advantage of being almost free of scale effects as the model scale generally approaches prototype scale. On the other hand, the laboratory effects, which are due to the fact that only the cross-shore transport is considered in a 2D wave flume, might be seen as a drawback. However, for the time scale and the wave conditions prevailing during extreme storm surges, this effect was found to be relatively small and within the range of other uncertainties (Newe, 2005). Rather than addressing scale and laboratory effects, the recommendations in Figure 4.12, which have been derived from the long experience using the GWK-flume in Hannover, are briefly discussed below. Further details are given in the PhD-thesis of Newe (2005) which also includes further references on the data and documents used to come up with these recommendations. As shown in Figure 4.12, a distinction is made between three phases for the recommendations: planning of the experiments, performance of the experiments, and analysis of the results.

Planning of the experiments

The specification of the initial profile, together with the availability of sufficient sediment in the model, represent the first important steps in the planning phase. For case

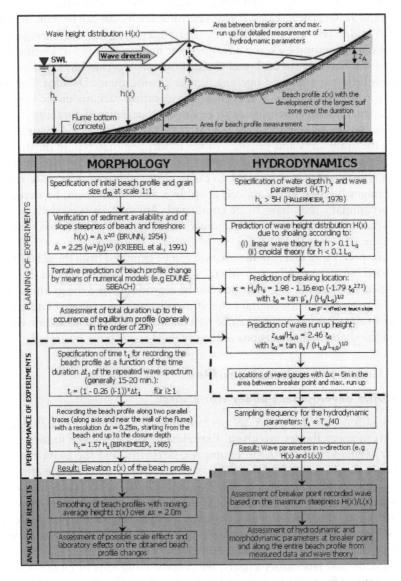

Figure 4.12 Recommendations for the planning, performance and analysis of large-scale experiments in wave flumes (after Newe, 2005).

studies, the actual beach profile and the sediment grain size D_{50} should possibly be reproduced at full or nearly full scale. It should be ensured – particularly in the case of basic research studies – that enough sediment is supplied in the model for an equilibrium beach profile to fully develop. Otherwise, too steep slopes of the seaward face of the reef and of the dry beach will develop, causing serious dissimilarities of the hydrodynamics and the cross-shore transport. A water depth h in the farfield (over the concrete bottom of the flume) larger than five wave heights ($h > 5H$) is required

(Hallermeier, 1981). Moreover, Bruun's (1954) rule and numerical modelling should be used the check the sediment availability.

The use of numerical models such as SBEACH and EDUNE allows a tentative prediction of the required test duration for the equilibrium profile to fully develop. For a sand beach with a grain size $D_{50} \approx 0.3$ mm, about 20 hours are generally required for experiments in the GWK. This approximate value was obtained from the analysis of the experimental data and corresponds to the time at which the offshore sediment transport induces no noticeable changes of the beach profile. However, a slight widening of the surf zone is still present, so that a fully static equilibrium profile can hardly be obtained in a wave flume (remaining nett transport between the waterline and the breakerline in the range of $q_t = 0.06$ m³/m · h). The tentatively predicted beach profile changes are very important to optimise the locations of wave gauges and further transducers to record the required hydrodynamic and morphodynamic parameters. Distances of $\Delta x \approx 5$ m between the wave gauges are generally sufficient in the GWK to assess the energy dissipation within the swash zone (Figure 4.12). The predicted beach profile changes are also very important for the elevation of other transducers to be installed along the beach profile (e.g. current meters).

For comparison, shoaling should be calculated using linear wave theory for larger water depth ($h > 0.1\ L_0$) and for smaller depth ($h < 0.1\ L_0$) cnoidal wave theory or the explicit approximation proposed by Muttray & Oumeraci (2000). The location of the breaker line can be assessed by the available breaking criteria such as those proposed by Goda (1970), Weggel (1972), etc., which provide a good approximation for regular waves with shorter periods ($T \leq 5$ s). For longer periods with $T \geq 10$ s, the breaker height is however underestimated so that the breaking criterion proposed by Newe (2005) in Figure 4.12 is more appropriate.

The decay of wave height after breaking may be estimated by a numerical model such as the one proposed by Thornton & Guza (1983) or simply by using empirically based formulae such as those proposed by Muttray & Oumeraci (2000) or the following formula by Andersen & Fredsoe (1983):

$$H(x) = 0.35H_b + 0.65\exp\left(-0.12\frac{x}{H_b}\right)H_b \tag{4.21}$$

where H_b is the breaker height and x_b the distance landwards from the breaker line.

For the prediction of the wave run-up height exceeded by only 2% of waves, z_{98}, as a function of the incident significant wave height in deepwater H_{s0} and the surf similarity parameter, the problem consists in the selection of the appropriate slope steepness $\tan \beta_s$ to be considered in the latter. Using the equivalent slope steepness proposed by Saville (1958), which represents an average value over the entire area between the breaker line and the maximum run-up, has led to the wave run-up formula shown in Figure 4.12 with a coefficient of variation of only 10% (Newe, 2005).

Performance of the experiments

A sampling rate of the water surface elevation and orbital flow in the range of 1/40 of the mean wave period T_m is sufficient (e.g. sampling rate = 8 Hz for $T_m = 5$ s in the model). Since the beach profile changes have to be recorded during each of the tests constituting the entire duration of the storm surge to be simulated and since the latter cannot be entirely simulated within a single test, time interval t_1 at which the records have to be performed should be optimized as a function of the duration Δt of each test. A relationship to assess time interval t_1 is proposed in Figure 4.12. On the other hand, time interval t_1 should also be adaptive in the sense that at the beginning of the experiments, i.e. when the cross shore transport rate is relatively high, the time interval should be kept much shorter ($t_1 \approx 15$–20 minutes) than at the end of the tests. After 15h test duration, a record every 2–3 hours will generally be sufficient. Moreover, a substantial time saving can be achieved, if a distinction is made between the areas of significant morphological changes and those with much less changes.

The records of the beach profile changes should be performed up to the closure depth $h_c \approx 1.56\ H_s$ (Birkemeier, 1985). In fact, no noticeable morphological changes in deeper areas were observed in previous experiments. A spatial resolution of $\Delta x = 0.25$ m along the middle of the wave flume and along another path at a distance between $\Delta y = 0.50$ m and $\Delta y = 1.50$ m (for a 5 m wide flume) was found to be appropriate.

Analysis of the results

From the hydrodynamic point of view the analysis should particularly focus on:

*   The development of the wave parameters $H(x)$ and $L(x)$ along the entire beach profile, including the identification of the shoaling, the breaking location and height, as well as the wave decay after breaking;
*   The wave run-up, including other processes in the swash zone.

Some problems are often encountered when trying to identify with enough accuracy the location of the breaker line from the analysis of the recorded wave steepnesses $H(x)/L(x)$. Visually observed locations were found to correspond with a probability of about 95% to those identified from the maximum wave steepness $[H(x)/L(x)]_{max}$ recorded along the beach profile.

Regarding the morphological data it might be convenient to remove the small ripples by smoothing the recorded beach profiles without affecting the larger morphological bed features. For this purpose a moving average of the recorded elevations over a distance of 2 m proved to be very appropriate. The increase of the water depth which would result is generally less than 2% (<3 cm).

4.10.3   Beach erosion – comparison of scaling laws

The scaling laws discussed in Section 4.4.2 are compared here with field and laboratory data-sets at multiple scales. Ito & Tsuchiya (1984, 1986, 1988) and Ito *et al.,* (1995) made detailed studies of beach erosion profiles in quasi-equilibrium conditions under low and high waves. They have used small-scale and large-scale physical models (undistorted) with regular waves and initial slopes of 1:15 and 1:30. Similitude between bed profiles of different scales is defined to exist when the difference is less than twice the experimental error (based on repeated tests). The low-wave cases show onshore transport with the formation of a swash bar at the upper end of the beach slope (1:30), while the high-wave cases show offshore transport with the formation of a breaker bar at the lower end of the beach slope (1:15). The prototype bed profiles are obtained from the results measured in large-scale wave flumes (offshore depth of 1 to 4.3 m; wave periods of 3 to 11 s). The scaling laws derived from these undistorted scale model (regular waves) series are given by Eqs. (4.9)–(4.11).

These scaling laws were applied to a storm-induced beach erosion event (14–18 March, 1981) on the Ogata coast facing the Sea of Japan. The significant wave height increased from 0.5 m to about 4 m in about one day, remained constant for the following day and decreased again after that. The storm-induced set-up was about 0.3 m. The beach sediments varied in the range of 0.2 mm in the offshore zone to about 0.4 mm at the beach. The depth scale was set to $N_h = 50$, the grain size scale was set to $N_{D50} = 3.7$ based on Equation 4.10. Two model sands were used to represent the beach material variation in the prototype. The wave height variation during the storm event was represented by schematizing it into three regular wave cases, each with constant but different wave height and period. The beach profile changes observed in the prototype were reproduced very well when the mean wave height was used as the representative wave height in the prototype and not quite as well when the significant wave height was used. Similar conclusions are given by Ito et al., (1995).

Hughes & Fowler (1990) performed small-scale model experiments (undistorted) aimed at reproducing large-scale experiments in the GWK; the latter being defined as the prototype. The prototype sediment is $D_{50} = 0.33$ mm and $w_s = 0.0447$ m/s. The applied scaling laws are:

$$N_{ws} = (N_h)^{0.5} \tag{4.22}$$

$$N_{Tm} = (N_h)^{0.5} \tag{4.23}$$

Given a depth scale of $N_h = 7.5$, the settling velocity scale is $N_{ws} = 2.73$ (model sand of 0.13 mm, $w_s = 0.0164$ m/s). In the prototype experiments, sand with a median diameter of 0.33 mm was placed in front of a concrete structure with a slope of 1 to 4. The initial sand slope was also 1 to 4. Both regular and irregular wave tests were done. In the case of regular waves the model erosion at the upper end of the beach was slightly (10%) under-estimated. Almost perfect agreement was obtained by increasing the wave height by about 10%. Comparing test results of irregular waves in the prototype and in the scale model, the agreement was found to be very good. Comparable profile development can be achieved between regular waves in the scale model and irregular waves in the prototype when the value of significant wave height is used as the regular wave height in the scale model (in contrast to Ito's result given above). Profile development is found to be approximately twice as fast in the scale model with regular waves. Hence, only half the number of waves is required to obtain the same quasi-equilibrium bed profile (at corresponding times). In all tests the (offshore) slope at the toe of the beach was much steeper in the prototype (0.33 mm) than in the scale model (0.15 mm). Probably, there was relatively large onshore transport at this location (shoaling waves) in the prototype (GWK, Hannover) causing a steeper slope.

The various grain size scaling laws are plotted in Figure 4.13. The scaling law of Ito & Tsuchiya (1984, 1986, 1988) deviates considerably from the other three scaling

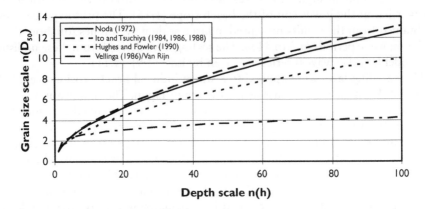

Figure 4.13 Grain size scale for sand as function of depth scale for various methods.

laws. Using the scale relationships, a physical scale model of a sandy beach/coast can be designed, as follows:

- Select depth scale and distortion scale;
- Find the grain size scale (or the fall velocity scale) from Eqs. 4.12 or 4.15, 4.7, 4.9, 4.10 and 4.22;
- Translate the measured results to prototype values using the applied depth scale (n_h and n_l) and distortion scale and the morphological time scale (n_{TM}).

Often, Eqs. 4.12 or 4.15 cannot be fully matched because of the given flume dimensions (flume not long enough to fit cross-shore profile in flume) and the available sediment material (finer or coarser than required). In that case *either* the morphological time scale could be adjusted, and the applied depth and length scales correspondingly, *or* use the length scale according to the scaling law (from Eqs. 4.12 or 4.15 instead of the applied length scale).

4.10.4   Scour and scour protection around monopiles

The presence of a monopile will cause scour in a non-cohesive seabed in cases where the current speed is greater than about half the current speed at the threshold of motion or where the Keulegan-Carpenter number, $KC = U_w.T/D \geq 6$, where U_w = near-bed wave orbital velocity, T = wave period and D = monopile diameter (Sumer & Fredsøe, 2002, Chapter 3). In these circumstances, the scour depth may become a critical feature of the monopile design and scour protection may have to be installed. The development of scour and the stability of scour protection can both be simulated in a physical model to assist in this process.

The Keulegan-Carpenter number represents the way in which the wave flow interacts with the monopile. Froude scaling preserves the Keulegan-Carpenter number for the monopile, which is sufficient to preserve similitude of the equilibrium scour depth and scour width for the case where the scour is in the live-bed regime. For clear-water scour, the Shields parameter should also be involved.

If the pile diameter is sufficiently large that diffraction effects become important ($D/L > O(0.1)$), then an additional parameter, the diffraction parameter, D/L, should also be involved. Here L = the wave length (Sumer & Fredsøe, 2002, Chapter 6).

For time scales or time development (of scour processes), the Shields parameter also needs to be involved (see Sumer & Fredsøe, 2002, pp. 206–212). The time scale of the scour process should be scaled by the quantity $D^2/[g(s-1)d^3]^{0.5}$ (Sumer & Fredsøe, 2002, p. 70) in which D = the pile diameter, g = the acceleration due to gravity, s = the specific gravity of sediment grains, d = the sediment size.

When modelling rock protection, both the frictional force and the pressure force should be considered. The friction force is determined by the Shields parameter, which can be calculated with the aid of the wave friction factor. Wave friction factor data are now available for very small values of KC (Dixen et al., 2008). The ratio of the pressure force to that of the friction depends on the KC number. Hence, if Froude scaling is applied to the rock protection also, the KC number of the rock will be scaled correctly and thus the pressure gradient term for the rock will also be scaled correctly.

Best practice for scour tests:

- Irregular waves;
- Waves and current, if current is significant;
- Check for the effect of water depth-to pile-diameter ratio, the effect of sediment gradation, the effect of cross-sectional shape (Sumer & Fredsøe, 2002, Chapter 3), and the effect of wave diffraction (Sumer & Fredsøe, 2002, Chapter 6);
- Measure time series of scour development in at least one case to establish the timescale of scour;
- Measure profiles before and after using a bed profiler / scanner;
- Keep a photographic record before and after each test.

Best practice for scour protection tests includes the above, plus the following:

- Scale rock to account for differences in density of water and rock between model scale and full scale;
- Preserve the same number of armour layers in model as in prototype;
- Typically use a template to generate the correct profile in constructing model;
- Use different colours sprayed onto armour to identify movements within and between different areas;
- Measure profiles before and after using a bed profiler / scanner;
- Photograph each scour protection system from the same fixed location, using the same camera and setting both before and after each test.

Post-installation surveys of some scour protection schemes have indicated that the placement of armour at full scale is a lot more haphazard than can be undertaken in the laboratory. The careful and fastidious placement of armour in the laboratory is likely to give a misleading, probably conservative, picture of the stability of armour placed at full scale.

Chapter 5

Ecological experiments

Lynne Frostick, Stuart McLelland & Theresa Mercer

Morten Alver, Leonardo Damiani, Ingrid Ellingses, Pam Naden, Alexandra Neyts, Lasse Olsen, Hocine Oumeraci, Ellis Penning, Ponnambalam Rameshwaran, Stephen Rice, Stefan Schimmels & Francesca de Serio

5.1 INTRODUCTION

Physical modelling of ecology is essential to improve our understanding of both the impact of environmental factors on biota and the impact of biota on their environment. Research on the interaction amongst water flow, morphology, sediment transport *and* biological processes is essential for better understanding of the natural environment and to improve management of the natural environment. Ecological experiments encompass a wide array of marine and freshwater studies, including pollutant dispersal and accumulation, biodiversity studies, bio-engineering, biomechanics, and erosion and deposition processes. Physical modelling in ecological experiments has the same advantages over field monitoring as hydraulic and sedimentological studies in that the hydraulic and sedimentological parameters can be controlled under experimental conditions to improve our understanding of the response of biota to environmental changes and for the development of better management strategies and practises. However, whilst physical models enable the investigation of the effect of hydraulic and sedimentological controls on ecology, experiments must be conducted under constraints that ensure the experimental environment sustains the biota being investigated which brings new constraints to the laboratory set-up. This field is rapidly expanding because of the pressures to improve and develop environmental management as well as academic interest and by its nature the field is inter-disciplinary involving biologists, ecologists and hydraulic engineers. Therefore it is timely to produce guidance for these research communities to enable a better understanding of the practicalities involved in ecological experiments and to guide practitioners towards best practice based on our current understanding.

Warren & Davis (1971) suggested that "artificial streams" offered a "better approach" to studying stream ecosystems than simple physical models or studies of natural stream systems. Warren and Davis suggested that laboratory stream ecosystems could be used to make observations on interacting individuals and populations and to obtain some understanding of the operation of communities, despite the "conceptual danger" in thinking that laboratory ecosystems were an inadequate representation of natural ecosystems. They emphasised the potential of artificial streams to help understand general concepts since they function as an ecosystem themselves.

Lamberti and Steinman (1993) edited a collection of 13 papers arising from a symposium entitled "Research in Artificial streams: Applications, Uses and Abuses". The aim of this work was to evaluate the progress of artificial stream research since the work of Warren and Davis (1971) and to evaluate whether it had provided a "better approach" to investigating stream ecosystems. It is helpful to emphasise the potential abuses of artificial stream experiments that were identified by Lamberti and Steinman (1993) which serve as a useful guide in terms of what we need to avoid when undertaking physical models of ecology:

- Using artificial streams strictly for experimental design purposes to increase the number of treatment replicates.
- Experiments that overwhelm or distort the ecosystem such that the manipulations produce a detectable response, but with little or no ecological relevance. Sometimes this may be useful for determining the threshold for response within an ecosystem, such as ecotoxicological studies.
- To evaluate natural phenomena from an experimental design that is very different from a natural system.
- To draw inferences about natural systems from experimental results without field validation to support the conclusions.

5.2   ECOLOGICAL PERSPECTIVE

Hydraulic and sedimentological processes are important at both the large scale and the small-scale for their influence on biota in aquatic environments and *vice versa*.

Vegetation may have a complex effect on flow roughness particularly since it can respond to the flow field itself, hence the roughness due to vegetation is variable due to plants being able to bend and change their shape and roughness can also be dynamic since the vegetation can move with a wave like motion which introduces a time-varying component to the roughness (e.g. Stephan & Gutknect, 2002). Water velocities under the canopy can be considerably lower than free-stream velocities, for example Gambi *et al.* (1990) measured velocity reduction of 2–10 times lower within the canopy compared to upstream of a seagrass bed. This low energy microenvironment can increase suspended sediment deposition and is important for benthic community structure (e.g. Peterson *et al.*, 1984). The hydraulic characteristics of flow within and around vegetation depends upon a number of factors such as the shoot density (e.g. Gambi *et al.*, 1990; Peterson *et al.*, 2004), shoot thickness (Bouma *et al.*, 2005), and patchiness (Folkard, 2005).

The flexibility of plants is important for their ability to survive and potentially dominate benthic environments where flow conditions would damage and/or erode rigid structures. Flexibility which enables plants to bend, change shape and/or alter their size can reduce the drag on a plant to reduce or resist the damage or erosion caused by extreme flow events such as floods, storm events and wave breaking (e.g. Boller & Carrington, 2007). Boller & Carrington (2007) identify two important aspects of reconfiguration, first that plants may change their frontal area thereby changing their total drag and second that plants may change their shape which changes their drag coefficient. Under waves, flexibility also enables plants to move with the flow which minimises the time that flow is accelerating past the plant thereby reducing the drag (e.g. Fonseca *et al.*, 1987).

The spatial configuration of plant stands may also respond to flow regimes, for example Fonseca *et al.* (2007) noted that seagrass shoots had been observed to align in rows perpendicular to the primary axis of the flow direction (under both unidirectional and oscillatory flow conditions). Such a structured linear arrangement of plant shoots may lead to reduced force and drag coefficients on shoots and improved light capture compared to a random arrangement of shoots on the sea bed (Fonseca et al., 2007).

The mass transfer of resources through processes such as photosynthesis and nutrient cycling for individuals and communities in plant stands respond to hydraulic factors such as flow velocity and Reynolds shear stress (e.g. Nishihara & Ackerman, 2006; Cornelisen & Thomas, 2006). For example, Cornelisen and Thomas (2006) show that in nitrogen and ammonium uptakes in seagrass communities increased with bulk velocity and Reynolds Shear stress at the top of the canopy. The structure of plant communities may also effect the spatial distribution of nutrient uptake as demonstrated by Morris et al. (2008) who found that nutrient uptake was 20% greater at the leading edge of seagrass beds and that different seagrass morphologies affect the rate of nutrient uptake due to the differences in within canopy water.

In both marine and freshwater systems, benthic organisms depend on hydraulic and sedimentological conditions for the supply of oxygen and food as well as the removal of waste products, therefore flow conditions at the grain/bedform scale and through the water column may be important for animal growth and development as well as the behaviour and spatio-temporal distribution of both individuals and populations or communities. For example, Frechette et al. (1989) have shown that the vertical distribution of phytoplankton over mussel beds is related to enhanced diffusive transport replacing the depleted near-bed zone and that the bed roughness due to the mussels themselves may enhance turbulent transport near the bed. Bed roughness due to mussel beds is an example of biogenic roughness which describes the surface roughness elements created by benthic organisms including structures like worm tubes, sand mounds and pits, snail shells, and resting crabs. The effect of such biogenic roughness is to alter the near bed flow structure (e.g. Huettel & Gust, 1992; Friedrichs & Graf, 2009) and the resulting patterns of flow acceleration and deceleration, flow separation and vortex development affect the ability of organisms to capture food (e.g. pathways and residence times for food particles), alter the exchange of gasses and potentially increase solute fluxes in surrounding permeable bed material.

Animal behaviour can also modify local flow patterns, for example Thomson et al. (2004) found that feeding S. vittatum larvae produce a downstream velocity reduction that extends at least 10–20 mm downstream which may explain reduced ingestion rates observed for downstream larvae and aggressive behaviour towards their upstream neighbours. Transported and deposited suspended sediment may affect organism growth rates due to the effect on nutrients (e.g. Peeters et al., 2006), ingestion rates due to the clogging feeding structures and reduction of feeding efficiency (e.g. Newcombe & MacDonald, 1991; Kent & Stelzer, 2008) and behavioural effects in response to the increase and/or decrease of sediment. These effects are important at the level of individuals, but they may and also affect populations, species diversity and richness and food web dynamics (e.g. Schofield et al., 2004; Rabeni et al., 2005; Stone et al., 2005). Feeding strategies (i.e. deposit feeding or suspension feeding) can also be influenced by both the flow velocity and suspended sediment concentration and for some species flow speeds and suspended sediment concentration can affect ingestion rates and/or growth rates (e.g. Finelli *et al.*, 2002; Hentschel & Larson, 2005).

Hydraulic conditions can also be important for the fertilisation processes, particularly where gamete movement is influenced by the flow conditions. For example, attenuation of flow within sea grass stands may be important for spatial variations in gamete fertilization (Peterson *et al.*, 2004). Flow velocities and vortex development have been shown to affect gamete advection and fertilization with gametes being dispersed in viscous fluids that cling to tests and spines (Yund & Meidel, 2003). Evidence from flume studies has shown that suspended sediment concentration may also impact on egg fertilization, with successful fertilization being reduced below 80% with suspended sediment concentrations in excess of 9000 mg/L for sockeye (*Oncorhynchus nerka*) and coho (*Oncorhynchus kisutch*) salmon gametes (Galbraith *et al.*, 2006).

Both hydraulic conditions and the sediment substrate are important factors for determining where larvae may settle. Flow velocities and turbulence intensities may directly influence animal behaviour, but are also important for the effect they will have on sediment erosion, transport and deposition in terms of changing the local environment and the response of animals to such changes. The movement of juvenile and adult benthic individuals may be either active or passive depending on species and hydraulic conditions.

5.3   HYDRAULIC ENGINEERING PERSPECTIVE

The interdisciplinary nature of ecological experiments highlights the important contribution that hydraulic engineers and physical modellers can bring to ecological experiments, particularly in terms of their understanding of the limitations and advantages of flume experiments. For example, recent work by Cooper *et al.* (2007) demonstrated that it is important to consider flume characteristics when investigating flexible macrophytes since the relative blockage effects associated with flume width may impact on the measured drag.

The interactions between biota and hydraulics described in section 5.2 above demonstrate the importance of biota for sediment dynamics in natural environments. Changes in the spatial distribution of shear stress produced by both plants and animals may be responsible for spatial and temporal variations in sediment transport and should not be neglected. There are also potential advantages for hydraulic engineering to exploit biogenic roughness attributes for ecological engineering.

5.3.1   Expectations of hydraulic engineers from collaboration with ecologists in laboratory experiments

This short note addresses four research topics related to biological effects in hydraulic/coastal engineering, including the damping effect of tree vegetation in flow and waves, the reinforcement of clay cover by grass to enhance dike stability, the effect of burrowing animals on dike stability and the bio-stabilization/bio-destabilization of sediment beds.

Damping effect of vegetation on flow and waves

Before starting with any hydraulic testing and/or mathematical/numerical modelling, the first step is to parameterize the considered vegetation from the point of view of hydraulic resistance (attenuation performance for flow and waves). The main goal is to determine a much simpler structure which is easy to model and which has an equivalent performance as its prototype counterpart.

In the case of coastal forest vegetation it is important with the collaboration of ecologists:

- To classify the tree species according to their location relative to the shoreline, their degree of submergence during flood events (fully/partially submerged), the structure and typical dimensions of their components;
- to define easily measurable/observable geometrical and biometrical parameters and accurately enough despite to the complex 3D-plant structure with randomly distributed trunks, branches etc;
- to determine a typical representative structure of an individual tree with all required parameters including the characteristic dimensions of all components (trunk, canopy, roots);
- to assess the distributions of the Leaf Area Index (LAI) over the plant height in the case of flexible leafy tree vegetation. Based on the self-similarity related to the stiffness of each tree species can at least be defined by four plant structure parameters: Branching ratio R_B, diameter ratio R_D, length ratio R_L and minimum diameter D_{min}. Moreover, the average trunk diameter D_{Trunk}, the average trunk length L_{Trunk} and tree height h_{plant} should be determined;
- to find for a given species the tree age as a relationship generally exists between age, length and diameter;
- to determine for a forest of trees of a given species, the typical density of trees, possibly including typical arrangements. Moreover, the height and density of typical undergrowth should also be determined as it strongly affects bottom friction and thus the overall hydraulic performance of the forest during floods.

Enhancement of erosion resistance of soil by vegetation (grass) roots

Testing the resistance of grassed dikes and levees against the action of water waves and overflow at full-scale under well-controlled conditions has become possible (availability of large-scale testing facilities) and is expecting to increase due to the risk associated with climate changes. Generally, the revetment of a dike is made of clay layer covered by grass. Both clay and grass builds a very complex composite material that needs to be better described from the erodibility point of view. In addition to a given clay category, the density (root volume ratio RVR) and depth of the roots are determinant.

The collaboration with ecologists is particularly required for the following tasks:

- Determination of most appropriate grass species providing the highest RVR;
- Assessment of biological degradation mechanisms under both intermittent and continuous hydraulic loading, including the capabilities to regenerate after degradation (between successive flood events);

- Lightening procedure and techniques for the grass cover over the entire testing durations in the laboratory (up to many months);
- Development of new techniques for an easy determination of the RVR and other grass vegetation parameters.

Effects of burrowing animals on the stability of sea dike and levees with grassed clay

Holes and cavities in sea dikes and levees caused by rodents such as common moles, rates, rabbits, muskrats etc. represent weak spots at which a breach of the flood defence can be initiated. Generally, a good clay revetment with a good grass cover can resist extreme flow and wave conditions without any significant damage to the defence structure. However, field evidence has repeatedly shown that this is not the case when such weak spots exist.

The availability of large-scale hydraulic testing facilities now allows us to study under well-controlled conditions the effect of such weak spots on the erosion of earthen hydraulic structures under extreme flow and wave conditions.

A closer collaboration with ecologists will allow us:

- To provide the typical location, geometry and dimensions of such weak points within the structure for each animal species, and then to proceed with a weak spot classification;
- to parameterize the weak spots in order to proceed with their simplified reproduction in the hydraulic model;
- to develop countermeasures for preventing such weak prints especially those occurring at the most critical locations.

Biological effects on the erosion resistance of sediment beds

In the last decades, considerable progress has been made on modelling morphological changes in river and coastal areas. However, only the physical processes have been so far included in the models while the biological effects which may be crucial for both sediment structure and sediment dynamics – and even hydraulics – have been largely ignored due to the high complexity of the related interactions and their modelling. Now, the advance of laboratory techniques has made possible that stabilization/destabilization processes can be studied under well-controlled conditions in flume experiments supplemented by field experiments to quantify the true impact under prototype conditions. A close collaboration with ecologists will allow us:

- A better and more systematically distinction between the biological activities with stabilization and those with destabilization as a net effect. This is very important for the hydraulic engineer to establish a prioritization of the effects to be investigated;
- to proceed with a systematic classification of the biological activities for those resulting in destabilization and those resulting in stabilization;
- to determine for each class and type of biological activity, the actual processes responsible for destabilization and those responsible for stabilization including

phenomenological descriptions, causalities and quantifications. This is important for the hydraulic engineer to decide on the feasibility of testing real organisms or surrogates in the laboratory;

- to improve and further develop the laboratory and modelling techniques to study bio-stabilization and bio-destabilization of sediment beds;
- to design and perform field experiments as a necessary supplement for the laboratory experiments to achieve a reliable quantification of the processes.

5.4   EXPERIMENTAL DESIGN

This section outlines the factors that should be considered when designing suitable physical models considering the formulation of the problem, scaling issues, dimensional analysis and the interaction between ecological and hydrological processes.

5.4.1   Dimensional analysis

Dimensional analysis, a simple and powerful analytical technique used in many fields of science, is based on the fundamental concepts (e.g. dimensional homogeneity) developed by Vaschy in 1882 and Rjabouchinskij 1911 and finally formalised by Buckingham (1914) in the so-called Pi-Theorem. In physical modelling dimensional analysis has been primarily used:

- To reduce the number and complexity of experimental variables that affect the physical processes under study;
- In planning, model testing, analysing and interpreting the experimental results;
- To derive scaling laws that allows testing models at a smaller scale than at prototype scale;
- To provide a better insight on controlling parameters and the nature of the complex phenomena being studied;
- To help derive equations for complex phenomena which cannot readily be derived from first principles;
- To check the consistency of equations, particularly those which are not derived from first principles ('check the units').

Dimensional analysis provides diverse and interesting applications in biological systems, including alleometry (e.g. Stahl, 1981, 1982; Günther & Morgado, 2003) as well as in ecology (e.g. Rosen, 1989). The scale of physical quantities common in biology such as the Vogel exponents, which scale the drag force with the flow velocity, were determined by sophisticated mathematical and experimental approaches (Alben *et al.*, 2002) and can be easily predicted with simple dimensional analysis (Gosselin *et al.*, 2009).

The premise of dimensional analysis is that the fundamental equations of motion are dimensionally homogeneous and that all relations derived from these equations must also be homogeneous. Therefore, dimensionally homogeneous equations constitute the basis for the applications of dimensional analysis. Specifically for physical modelling, dimensional analysis is conducted in five major steps (Oumeraci, 1994b) which are briefly summarised below. For more details refer to easily accessible textbooks (e.g. Hughes, 1993a).

Step 1: Identify and select independent variables

For each variable x_1 we are looking for, the complete set of important independent variables $(x_2, x_3, \ldots x_n)$ that determine the dimensionally homogeneous equation $x_1 = f(x_2, x_3, \ldots x_n)$. This represents the most important and crucial step as it requires a very good insight into the physical processes being studied, including the basic laws that govern these processes, and it should result in a correct compromise between too many variables (increase in complexity and testing efforts) and insufficient variables (incorrect results and conclusions). Pertinent x-variables in most engineering problems are related to geometry, material properties and external effects that induce changes in the physical system. Moreover, time and physical constants such as acceleration due to gravity may be useful in forming dimensionless Pi-variables in Step 3. It is also important to make sure that all selected variables are independent. For instance, fluid density is related to fluid specific gravity by acceleration due to gravity, so that only two of these three variables can be considered as independent.

Step 2: Determine dimensions of all variables and build dimensions matrix

Each dependent variables and related independent variables are expressed in terms of their respective r-basic dimensions (e.g. r = 3 for l, t, m being the dimensions for length (m), time (s) and mass (kg), respectively): $(x_1) = l_{a11} \, t_{a21} \, m_{a31}$, $(x_2) = l_{a12} \, t_{a22} \, m_{a32}$, ..., $(x_n) = l_{a1n} \, t_{a2n} \, m_{a3n}$ where the exponents aij are dimensionless numbers that follow from each quantity's definition (indices i and j indicate the row and column of the dimension matrix (ltm-matrix, respectively).

Step 3: Determine number of dimensionless Pi-variables and equations set for k-exponents

The number of Pi-variables is obtained by applying the Pi-theorem which tells us that in a dimensionally homogeneous equation involving 'n' variables, the number of dimensionless variables (Pi-variables) that can be formed is decreased to 'n-r' where 'r' is the number of fundamental dimensions encompassed by the variables. The 'n-r' Pi-variables are obtained by building a dimensionless product of the variables $x_1 \ldots x_n$ with exponents $k^1 \ldots k^n$ expressed in terms of dimensions $(Pi) = (x_1)^{k1} \cdot (x_2)^{k2} \ldots (x_n)^{kn}$ (with (Pi) = ltm = 1). This product together with the ltm – atrix from step 2 provides a set of 'r' linear equations which must be solved to determine exponents $k_i \ldots k_n$ (see step 4). If for instance n = 6 and r = 3, then a system of three equations with six unknowns $k_1 \ldots k_6$ would result. It might be useful to notice that the Pi-theorem tell us neither the forms of the dimensionless variables $Pi_1, Pi_2, \ldots, Pi_r$ nor the form of the dimensionless relationship: $Pi_1 = g(Pi_2, \ldots, Pi_n)$.

Step 4: Solve equations set for k-exponents and determine dimensionless Pi-variables

For this purpose the 'n' – unknown exponents $k_1 \ldots k_n$ in the system of the 'r' – equations formulated in Step 3 must be determined. Since this system of equations is indeterminate, 'n-r-k – exponents must be set before solving the equations system. Considering the example n = 6 and r = 3, a set of three equations with 6 unknown is obtained, so that three unknowns must first be specified before solving the system of equations. No generally accepted rules are yet available on how to meaningfully specify/select these unknown. Nevertheless, the following considerations might be useful in the selection/specification of the k – exponent: (i) each controllable variable should possibly occur only in one Pi-variable, (ii) the dependent variable should not occur in more than one Pi-variable, (iii) one should seek to possibly obtain dimensionless Pi-variables which correspond to "standard" fluid flow dimensionless numbers such as Reynolds (*Re*) and Froude (Fr) number (Table 5.1). Re- and Fr-numbers are more commonly used in standard hydraulic engineering models where generally stiff bodies and structures are considered (e.g. Oumeraci, 1984; Hughes, 1993a). Introducing biology and ecology in hydraulic modelling, where flexible and movable bodies are more common, will result in putting more focus on the effect of flexibility (Cauchy number), the effect of surface tension (Weber number), flow oscillations (Strouhal- or Keulegan-Carpenter number KC-number) as well as on useful numbers describing the interaction (coupling) between flow and body such as the reduced frequency f_R (Table 5.1).

Table 5.1 Dimensionless numbers for fluid flow, structure and flow-structure-interaction (in this instance L denotes length).

		Parameter	Definition	Remarks
Fluid flow	1	REYNOLDS-number	$Re = \dfrac{U \cdot L}{\nu}$	Time scale ratio of viscous diffusion an convection
	2	FROUDE-number	$F_R = \dfrac{U}{\sqrt{g \cdot L}}$	Time scale ratio of gravity wave propagation and fluid flow
	3	MACH-number	$M = \dfrac{U}{c}$	Time scale ratio of sound wave propagation and fluid flow
	4	STROUHAL-number	$St = \dfrac{L}{U \cdot T}$	Time scale ratio of mean flow and oscillatory flow (used in an unidirectional flow)
	5	KEULEGAN-CARPENTER-number	$KC = \dfrac{U \cdot T}{L}$	Time scale ratio of Oscillatory flow to mean flow (used in water waves)
Structure	6	Relative structure motion	$\mathcal{D} = \dfrac{\xi}{L}$	Ratio of structure motion and characteristic length of structure
	7	Deformation number	$\mathcal{G} = \dfrac{\rho_s \cdot g \cdot L}{E_s}$	ps = mean density of structure Es = Elasticity modulus of structure
Fluid structure interaction	8	Mass number	$\mathcal{M} = \rho_F \cdot \rho_s$	Ratio of fluid density and structure density
	9	Reduced frequency	$f_R = \dfrac{f_e \cdot L^2}{U} = \dfrac{1}{F_R \cdot \sqrt{G}}$	Time scale ratio of flow around structure and structure motion
	10	STOKES-number	$S_T = \dfrac{f \cdot L^2}{\nu} = R_E \cdot f_R$	Time scale ratio of viscous diffusion and structure motion
	11	Dynamic FROUDE-number	$F_D = \sqrt{\dfrac{f^2 \cdot L}{g}} = F_R \cdot f_R$	Time scale ratio of gravity wave propagation and structure motion
	12	Compressibility number	$\mathcal{K} = \dfrac{f \cdot L}{c} = M \cdot f_R$	Time scale ratio of sound wave propagation and structure motion
	13	CAUCHY-number	$C_y = \dfrac{\rho_F \cdot U^2}{E} = \dfrac{\mathcal{M}}{f_R^2}$	Measure for elastic deformation of structure by dynamic pressure

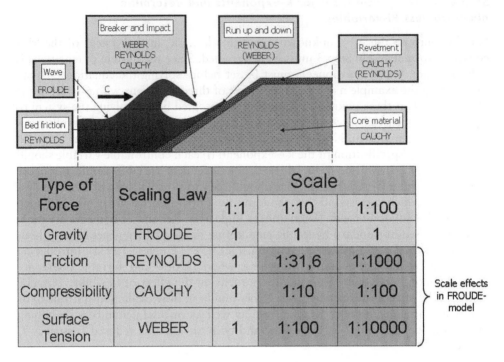

Figure 5.1 Scale effects in a Froude model (modified from Führböter in Oumeraci, 1994b).

Further important dimensional numbers for scaling the drag of moving/flexible bodies have recently been reviewed for instance by Zhu (2008) and Gosselin et al. (2009) and those to describe the interaction between flow and fishlike moving bodies by Triantafyllou et al. (2005). Reference can also be made to other reviews of dimensional analysis in biology (e.g. Stahl, 1981 & 1982; Günther & Morgado, 2003) and in ecology (e.g. Rosen, 1989).

It should be stressed that the selection of the most appropriate candidates from those and further physically meaningful dimensionless numbers always requires a very good insight into the physical processes involved in the problem. In coastal engineering, for instance, the following additional dimensionless numbers might be useful: dispersion parameter h/L (h = water depth, L = wave length), wave steepness H/L (H = wave height) or wave steepness parameter $H/(gT^2)$ (T = wave period), breaker index h/H or surf similarity parameter $\tan\alpha/\sqrt{(H/L)}$ ($\tan\alpha$ = bed slope angle), Dean parameter $H/(w_sT)$ (w_s = sink velocity of sediment in water), Keulegan-Carpenter number KC, etc. Moreover, the ratio of water density to sediment density, grain size related Reynolds number, the Shields number (as a dimensional bed shear stress), the ratio of bed shear stress to critical bed stress, etc. may also represent relevant dimensionless numbers where sediment transport and seabed scour processes are involved.

The ultimate result of this step is the formulation of the dimensionless Pi-numbers Pi_1, Pi_2, ..., Pi_r as a function of the variable x_1, x_2, ..., x_n as defined in

Step 1, so that the experiments can now be planned to obtain the relationship $Pi = g$ $(Pi_1, Pi_2, ..., Pi_r)$.

Step 5: Verification

Every variable $x_1, ..., x_n$ identified in Step 1 should occur in at least one of the Pi-variables which were formed in Step 4. The variable x_1 we are looking for, must occur only in one of these Pi-variables. Fractional k – exponents which might result in a Pi-variable should be avoided by transforming the Pi-variable. Transformation may also be beneficial for isolating certain parameters and enabling better experimental control. However, the transformation must result in the same number of new Pi-variables as originally determined in step 3 (e.g. Langhaar 1951, for transformation techniques). Ideally, an attempt should also be made to justify/interpret the final relationship $Pi = g$ $(Pi_1, ..., Pi_r)$ through mathematical analysis and physical reasoning.

5.4.2 Physical modelling and scaling considerations

Based on the results of the dimensional analysis the functional relationship between the Pi-variable we are looking for, and the other Pi-variables can be determined by mathematical and physical modelling. However, in most cases involving both biology and hydraulics, the problem becomes so complex and so difficult (3D complex structures, non-linear processes and interactions, non-steady state kinematics, oscillatory flow and turbulence, gaps in the understanding of associated biophysical processes, etc.) that the development of a useful mathematical model is rarely feasible. In such cases, physical modelling not only offers several advantages (faster than mathematical model development, enhancement of intuitive understanding of biophysical processes and insight in underlying mechanisms and interactions), but is often a must, particularly, when the knowledge gaps related to the biophysical processes involved are important.

The usefulness, advantages, limitations and examples of physical modelling in biomechanics is provided in many textbooks (e.g. Vogel, 1994; Biewener, 1992) and other review papers (e.g. Triantafyllou *et al.*, 2005; Koehl, 2003). Further reviews emphasize the role of biophysical modelling in the advance of flow and organism interaction allowing the construction of better organism inspired devices, robotic vehicles and systems (e.g. Lauder & Madden, 2006).

The necessity of observations and measurements in nature to provide the relevant data/information (size, shape, material properties, spatial and temporal distribution of hydraulic loads on plants and organisms, including changes during ontogeny, habitats, activities and ecological role, etc) required for the planning and execution of biophysical experiments has been underlined in many publications. The importance of coupling the knowledge of the environment, the ecological roles of the individual plants/organisms and life history strategies to gain more useful biophysical insight for laboratory testing has also been stressed (e.g. Koehl, 1999 for organisms and Langre, 2008 for plants).

When biophysical aspects are introduced in the physical model as commonly applied by hydraulic engineers, several effects and problems additionally arise which are related to scaling and a variety of further aspects.

5.4.3   Implications for biophysical modelling

When biophysical aspects are also introduced to a model, several classes of problems and effects make physical modelling and scaling as commonly practiced by hydraulic engineers much more complicated. From the biomechanical viewpoint, most complications arise from the fact that the fluid flow strongly interacts with (i) flexible plants which bend to reduce their drag and (ii) movable organisms which actively respond and adapt to the flow in order to survive, to feed, to 'fertilize' or to achieve an optimal motion (locomotion, swimming, etc.). In fact, most of the models in traditional hydraulic engineering include rather rigid structures or structures which respond passively to the flow. Hence, both construction and scaling of physical models become more complex and more difficult.

Conflicting scaling requirements and compromise

As mentioned above full dynamic similarity cannot be achieved in a single model. Conflicting requirements from the different scaling laws means that a proper scaling of all performance aspects cannot be achieved. Hence, a compromise between the different aspects of model performance that need to be reproduced must be achieved before model construction.

As a prerequisite, a very good biophysical and ecological insight in the organisms being studied, their ontogenic scaling and their interactions with the flow they inhabit would be required. Useful information on the dimensionless numbers that must be maintained for the proper scaling of different processes and performance aspects can be found in textbooks for fluid mechanics, biomechanics, materials science, hydraulic and structural engineering. Further useful information on ontogenic scaling, allometry, life history strategies and role in ecology of the organisms being studied can be provided by biology and ecology textbooks. Depending on the nature, size and complexity of the organisms and processes to be reproduced in a physical model, different approaches to dimensional analysis can additionally be used to achieve a proper scaling.

The most obvious – yet also most laborious – alternative is to reproduce differently scaled models of the same prototype counterparts in nature, but designed to study different performance aspects requiring different scale numbers to be maintained, so that the assumptions underlying the compromised scaling and their implications can be verified (e.g. drag with stiff and with flexible assumption, inception of motion and transport by flow, etc.). A further laborious alternative is to make use of a model family (i.e. the same model at different scales up to near full-scale) in order to better identify and even quantify possible scale effects in modelling the different processes.

A further approach is to use mathematical modelling in combination with physical modelling in order to quantify possible scale effects or to study parts of the performance aspects which cannot be simulated properly in the physical model. This may be particularly useful where the related processes cannot be measured/observed properly in the laboratory (e.g. diffusion of micro particles carried in suspension by turbulent flow – very important for organism feeding – can be calculated by using the measured flow field) or when their scaling is conflicting significantly with the scaling of other important processes. (e.g. Re-number for flow viscosity and Peclet-number for mass diffusion are both important but impossible to maintain simultaneously).

Regarding the **interaction of flow with compliant structures in a fixed bed model**, a better use should be made of the knowledge and experience available in structural dynamic engineering which are related to combined CFD and CSD modelling of the interaction between flow and deformable/movable structures and bodies. In addition to the dimensionless numbers related to the flow as commonly used in hydraulic engineering, further numbers related to the deformation/motion of the structure as well as to both flow and structure may be used for dimensional analysis and scaling.

Based on these numbers and their relative physical importance for the problem being studied, a more focused strategy to come up with the proper assumptions/simplifications for both physical modelling and CFD-CSD modelling could be developed. The validity of these assumptions/simplifications, including the scale of the physical model which will be adopted, should finally be verified.

Besides the approaches mentioned above, further alternatives might also be used. The most obvious alternative is a direct and systematic comparison with observations/measurements available from the field. Another way is to perform a systematic sensitivity analysis of the assumed model features/behaviour and the effect of their variation on the performance aspects to be obtained from the physical model. For instance, varying the stiffness of the model and measuring the effects on the flow and the resulting forces and structure responses will not only help to verify a 'rigid model assumption', but also quantify the implications of such an assumption on the prospective results. The sensitivity analysis may be significantly improved by also making use of all the modelling tools available.

Regarding the interaction of flow with sediment in a movable bed model, including incipient motion and different sediment transport modes (bed load, saltation, suspended load, sheet flow) the problem becomes even more complicated. Numerous attempts using dimensional analysis to define similarity criteria have resulted in diverse additional similarity numbers, depending on the specific problem addressed (see chapter 4). All these similarity parameters should be the same in model and in prototype to achieve a proper dynamic similarity (see figure 5.2). This, however can be satisfied only at a scale of 1:1. Generally, it is the inability to correctly reproduce the grain size D in smaller scaled model (D for non-cohesive sediment should not be less than 0.08 mm to avoid adverse cohesion effects) which introduces most of the problems, and thus scale effects:

- Mobility number F^*: is also called the Shields parameter or densimetric Froude number and represents the ratio of the disturbing forces (shear stress $\rho_w v^{*2}$) to the restoring force (sediment mass density in water $(\rho_s - \rho_w) D$). This is the most important similarity parameter as it describes both the initiation of motion and the transport of the sediment by the flow. Therefore both incipient motion and transport will be adversely affected if the similarity of F^* is not preserved. In this case, the wave-induced flow or other currents generally take longer to move the sediment resulting, resulting in slower erosion in the model. As the amount of suspended load is closely related to F^*, less suspension load would generally occur in the model. Though scale effects are never due only to one type of dissimilarity, the mobility number is often the most significant scale effect.
- Grain size Reynolds number: is a modified Reynolds number (Re) in which the characteristic length is represented by the grain size D of the sediment and the

characteristic velocity by the shear velocity v^* (related to bottom shear stress τ_b by the fluid density ρ_w: $v^* = \rho_w \cdot \tau_b$) of the near-bed flow (bottom boundary layer). When the viscous effects are negligibly small in the model (e.g. highly turbulent boundary layer, dominating suspended load), the resulting scale effects are generally of much less concern.

- Relative length l/D: represents the length scale l of the flow (e.g water depth h for unidirectional flow, wave amplitude for wave-induced flow) related to the sediment grain size D. Dissimilarity generally results in exaggerated forces required to move the sediments which may lead to higher transport rates Q in the model. Moreover, incorrect representation of the bedforms, bed friction and percolation would result.

- Relative density ρ_s/ρ_w: describes the density of the transporting fluid ρ_w relative to the sediment density ρ_s. Generally, this does not present any difficulty as long as the same sediment material is used in the model. However, if light-weight material is used, significant scale effects will be expected (alteration of transport mode, dissimilarity in soil liquefaction, piling up of sediment where not submerged due to relatively heavier particles in air, more energy dissipation due to higher porosity). Preservation of this parameter is particularly important in suspension-dominated transport models since it results in the dissimilarity of the relative fall velocity which is a determinant factor for suspended load.

- Relative fall velocity v/w: represents the importance of the flow velocity v relative to the fall velocity w of the sediment, which in coastal engineering is referred to as the Dean parameter $D_w = H/(w_s T)$ (H, T: wave height and wave period). For suspension-dominated transport models, preservation of this parameter together with the relative density is critical.

One of the most important implications from these considerations is that scaling of a physical movable bed model must be examined on a case-by-case basis. Therefore, the most important step is to identify systematically the most dominant processes affecting the performance aspect being studied and the associated influencing parameters. For instance, in the surf zone suspension load dominates whereas further offshore bed load or sheet flow dominates. Further aspects which make the scaling even more complicated is air entrainment which is particularly common in highly turbulent flow and suspension dominated transport (e.g. surf zone). As fresh water is used in the model, larger air bubbles than in salt water would result. This will affect both air content and residence time in the water column, and thus also the suspended sediment and other particles/organisms. For more details reference may be made to Hughes (1993a), Oumeraci (1984, 1994) and chapter 4.

Further considerations

Laboratory tests aiming at the study of interactions between flow of different regimes and types (e.g. laminar and turbulent, steady and unsteady, unidirectional and oscillatory) and deformable bodies with various features (e.g. size, shape, orientation to flow, surface roughness and coating, buoyancy and flexural stiffness, group arrangement) fixed to different substrata (e.g. cohesive and non-cohesive, rocky, etc) or moving in the ambient flow are typical in bio-ecological experiments. Therefore, such experiments are not only more difficult to scale, but also more difficult to perform and to analyse. Complications may arise from diverse sources, which may include: (i) the complex effect of morphology and flexibility on the flow-induced forces; (ii) the implications of low model Reynolds-numbers on the physical model; and (iii) the effect of turbulence on feeding, fertilization, survival and other activities of organisms.

Flexibility and morphology effects: Laboratory experiments that reproduce the interactions between flow and organisms are often primarily aimed at the determination of the drag they experience in the ambient flow. In such cases, it is important for both scaling and measuring to know that the experienced drag force is not proportional to the square of the velocity as predicted by classical theory for rigid bodies. In fact, flexible bodies adapt their shape as a function of the velocity. This reconfiguration is typical and beneficial for many organisms as this may result in a very substantial drag reduction. The issue of drag and reconfiguration of compliant systems may have vital implications in ecology (e.g. for plants to adapt to their habitats, to predict and prevent uprooting and breakage), as well as in structural engineering (e.g. paradigm shift to adaptive structures taking advantage of flexibility). Fundamental knowledge generated in recent years (Alben *et al.*, 2002; Bejan, 2005; Zhu, 2008; Langre, 2008; Gosselin *et al.*, 2009) may help to provide the theoretical interpretation that will unify the mechanics of flow and deformable bodies, and physical models in structural/hydraulic engineering and ecology. Innovative experiments which may lead to isolate the role of specific features of components and material properties of the deformable body on its dynamic response would be required to produce generic results.

For instance, soap-film tunnel experiments to describe 2D-hydrodynamics have enabled the isolation of the role of elastic bending of a flexible fibre in reducing the drag and determining a scaling law for drag as a function of the velocity for high Reynolds numbers $Re = 2000–40000$; i.e. drag is proportional to $v^{4/3}$ instead of v^2 for rigid bodies, implying that the so-called Vogel exponent e in $F_d \sim v^{2+e}$ is equal to $e = -2/3$ (Alben *et al.*, 2002). Despite the limitations of this law which have been demonstrated by more recent studies, this type of simple model has provided a generic insight into the self-similar behaviour of a flexible body as it becomes increasingly deformed. Zhu (2008) extended this study to lower Re-numbers by numerical modelling of incompressible viscous flow, showing that the scaling varies with the Re-number and another Cauchy-like dimensionless number which describes the relative importance of fluid kinetic energy and body elastic (potential) energy. A further extension of this study by Gosselin *et al.* (2009), using systematic testing in a wind tunnel, and dimensional analysis and theory to determine the dimensionless numbers for a proper modelling of the reconfiguration processes, has brought more insight into the mechanisms and difficulties involved. However, no final conclusions can yet be drawn on the proper scaling. Using physical models to determine the mechanical implications of structural features and material properties of the organisms or/and their components is also difficult, because the required information from the field are often unavailable and very hard to obtain. Moreover, organisms may actively change their size and other features unlike surrogates used in the model. This emphasises the potential limitations of physical modelling.

Low Reynolds number flow: For very small organisms moving in water at very low Re-number, near field flow and forces are difficult to measure and to observe

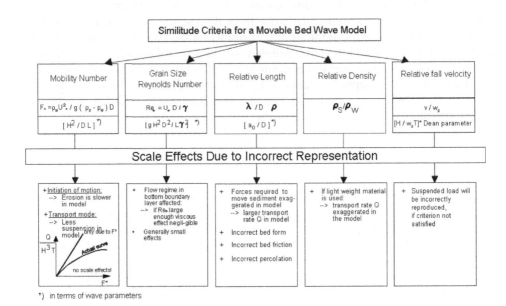

Figure 5.2 Similarity numbers and scale effects in sediment transport models (modified from Oumeraci, 1994a).

precisely. In this case, models that are larger than in nature are often used. To maintain the same Re-number in both model and prototype ($Re = v\,L_s/\mu_v$) the larger surrogate (L_s) has to move even more slowly (v) or a fluid with a higher viscosity (μ_v) would be required. Moreover, using larger models, other forces (e.g. surface tension) which are relevant for the behaviour of small organisms may become irrelevant for the larger surrogates.

A further implication of using a larger model is that the flow in the near field is affected many model lengths away from the model, so that wall effects in the testing tank may become crucial for the flow-induced forces. Further tests to quantify such effects might be required (e.g. Loudon *et al.*, 1994; Kumar & Anantha, 2007).

5.4.4   Turbulence

Turbulence is generated either by mean shear (loss of mean-flow kinetic energy) or by unstable stratification (loss of potential energy), resulting in the creation of eddies, which grow and eventually dissipate. Therefore, turbulence is mainly produced where water shears over a solid surface or against another water body with a very different speed. The first type occurs along solid walls, i.e. shores and bottoms in the case of the marine environment. The second type is found along sharp density interfaces or cavity-like flow (i.e. a stream along a body of stagnant water) or flow over a strongly varying bottom topography. The majority of examples of the second type of turbulence generation occur primarily in the horizontal plane (i.e. this case is defined quasi-2D turbulence) and can generate eddies significantly larger than the water depth.

In the coastal zone, mean shear is typically generated by winds or tides, but also by surface waves (Stokes drift) and baroclinic flows, including nonlinear internal waves. Unstable stratification results from surface processes such as surface cooling, evaporation or freezing and also as a result of differential advection. In the latter, vertically well-mixed waters become unstably stratified through current shear in situations such as upwelling zones, tidal estuaries and coastal areas during flood. Destruction of turbulence occurs by transformation into potential energy (during stable stratification) or viscous dissipation into heat (Burchard et al., 2008).

Turbulence therefore depends upon many effects, such as wall roughness or fluctuations in the inlet stream, but the primary parameter is the Reynolds number Re. The reason is that a profound change in fluid behaviour occurs at moderate Reynolds numbers. The flow ceases being smooth and steady (laminar) and becomes fluctuating and agitated (turbulent). The changeover with increasing Reynolds number is called the transition to turbulence. The flow will appear steady on average but will reveal rapid and random fluctuations if turbulence is present. If the flow is laminar, there may be occasional natural disturbances which damp out quickly. At the transition to turbulence, there will be sharp bursts of turbulent fluctuations as the increasing Reynolds number causes a breakdown or instability of laminar motion. At sufficiently large Re, the flow will fluctuate continually and is termed fully turbulent. The fluctuations, typically ranging from 1 to 20% of the average velocity, are not strictly periodic but are random and encompass a continuous range, or spectrum, of frequencies (White, 1994).

In flows which are originally laminar, turbulence arises from instabilities at large Reynolds numbers. Laminar pipe flow becomes turbulent at a Reynolds number (based on mean velocity and diameter) in the region of 2000 unless great care is taken to avoid creating small disturbances that might trigger transition from laminar to turbulent flow. Boundary layers in zero pressure gradient become unstable at a Reynolds number $V\delta^*/v_k = 600$ approximately (δ^* is the displacement thickness, V is the free-stream velocity, and v_k is the kinematic viscosity). Free shear flows, such as the flow in a mixing layer, become unstable at very low Reynolds numbers because of an inviscid instability mechanism that does not operate in boundary-layer and pipe flow (Tennekes & Lumley, 1972).

The principal characteristics of turbulent flows can be summed up in the following way (Tennekes & Lumley, 1972):

- Irregularity or randomness, which makes a deterministic approach to turbulence problems impossible; instead, statistical methods are used;
- Diffusivity, which causes rapid mixing and increased rates of transfer of momentum, kinetic energy, and contaminants such as heat, particles, and moisture. The rates of transfer and mixing are several orders of magnitude greater than the rates due to molecular diffusion;
- 3D vorticity fluctuations;
- Dissipation, due to viscous shear stresses that perform deformation work which increases the internal energy of the fluid at the expense of the kinetic energy of the turbulence. Turbulence needs a continuous supply of energy to make up for these viscous losses. If no energy is supplied, turbulence decays rapidly;

- Continuum phenomenon, therefore turbulence is governed by the equations of fluid mechanics;
- Turbulence is not a feature of fluids but of fluid flows.

No general approach to the solution of problems in turbulence exists. The equations of motion have been analyzed in great detail, but it is still next to impossible to make accurate quantitative predictions without relying heavily on empirical data. Statistical studies of the equations of motion always lead to a situation in which there are more unknowns than equations. This is called the closure problem of turbulence theory: one has to make *ad hoc* assumptions to make the number of equations equal to the number of unknowns (Tennekes & Lumley, 1972; White, 1998).

Turbulence consists of random velocity fluctuations, so that it must be treated with statistical methods, by means of a simple decomposition of all quantities into mean values and fluctuations with zero mean. We shall find that turbulent velocity fluctuations can generate large momentum fluxes between different parts of a flow. A momentum flux can be thought of as a stress; turbulent momentum fluxes are commonly called Reynolds stresses. The momentum exchange mechanism superficially resembles molecular transport of momentum. The latter gives rise to the viscosity of a fluid; by analogy, the turbulent momentum exchange is often represented by an eddy viscosity.

In fact, starting from the Navier-Stokes equation, we can write:

$$\frac{\partial u_i}{\partial t} + u_j \frac{\partial u_i}{\partial x_j} = -\frac{1}{\rho}\frac{\partial p}{\partial x_i} + v\frac{\partial^2 u_j}{\partial x_i \partial x_j} \tag{5.1}$$

where x_i corresponds to the axes of a Cartesian system, with $i, j = 1, 2, 3$ and u_i (with $i = 1, 2, 3$) are the three velocity components along these axes.

The velocity u_i (with $i = 1, 2, 3$) is decomposed into a mean flow U_i (capital letters are used for mean values) and velocity fluctuations u_i' such that:

$$u_i = U_i + u_i' \tag{5.2}$$

We interpret U_i as the following time average:

$$U_i = \lim_{T \to \infty} \frac{1}{T} \int_{t_0}^{t_0+T} u_i dt \tag{5.3}$$

Whose mean value is zero by definition.

The continuity equation for the turbulent velocity fluctuations yields to:

$$\overline{u_j' \frac{\partial u_i'}{\partial x_j}} = \frac{\partial}{\partial x_j}\overline{u_i' u_j'} \tag{5.4}$$

where the first term is analogous to the convection term $U_j \partial U_i / \partial x_j$ and it represents the mean transport of fluctuating momentum by turbulent velocity fluctuations.

By means of this equation, the equations of motion for the mean flow U_i, after some rearrangements, yields to the Reynolds momentum equation:

$$U_j \frac{\partial U_i}{\partial x_j} = \frac{1}{\rho} \frac{\partial}{\partial x_i} \left(\Sigma_{ij} - \rho \overline{u_i' u_j'} \right) \tag{5.5}$$

where the over-bar indicates time averages and Σ_{ij} is the mean stress tensor, given by

$$\Sigma_{ij} = -P\delta_{ij} + \mu \left(\frac{\partial U_i}{\partial x_j} + \frac{\partial U_j}{\partial x_i} \right). \tag{5.6}$$

Analogously the stress fluctuations σ_{ij} are given by:

$$\sigma_{ij} = -p'\delta_{ij} + \mu \left(\frac{\partial u_i'}{\partial x_j} + \frac{\partial u_j'}{\partial x_i} \right) \tag{5.7}$$

with P mean term of pressure, p' pressure fluctuation and μ molecular viscosity.
   The total mean stress in a turbulent flow may be written as:

$$T_{ij} = \Sigma_{ij} - \rho \overline{u_i' u_j'} = -P\delta_{ij} + \mu \left(\frac{\partial U_i}{\partial x_j} + \frac{\partial U_j}{\partial x_i} \right) - \rho \overline{u_i' u_j'} \tag{5.8}$$

and the contribution of the turbulent motion to the mean stress tensor is designated by the symbol $\tau_{ij} = -\rho \overline{u_i' u_j'}$, called the Reynolds stress tensor. The Reynolds stress is symmetric and the diagonal components of τ_{ij} are normal stresses (pressures). In many flows, these normal stresses contribute little to the transport of mean momentum. The off-diagonal components of τ_{ij} are shear stresses and they play a dominant role in the theory of mean momentum transfer by turbulent motion.
   The decomposition of the flow into a mean flow and turbulent velocity fluctuations isolated the effects of fluctuations on the mean flow. However, the equations for the mean flow contain the nine components of τ_{ij} (of which only six are independent of each other) as unknowns additional to P and the three components of U_i. This illustrates the closure problem of turbulence (Tennekes and Lumley, 1972). It should be observed that turbulence transports passive contaminants such as heat, chemical species and particles in much the same way as momentum.

Effect of turbulence on biophysical modelling

Turbulence interacts with organisms and their habitat, and may affect their activities and mechanisms in diverse ways, including for instance dispersion of suspended matter (feeding, etc.) and dislodgment of organisms from substratum. In this respect, basic research is considerably lagging behind most of other research areas. Generally, the dispersion of dissolved or suspended matter depends on both the turbulent and mean flow properties of the fluid. Small-scale flow conditions (range of cm or less) observed in the field, although difficult to reproduce quantitatively and to measure directly in the scaled model, can be visualised (e.g. video records of laser illuminated slices of the flow field to map the dye concentration). Moreover, the drag force on organisms exposed to turbulent flow fluctuates around a mean, and it is just this mean which is commonly cited to characterise the drag. Important information may be lost if turbulent fluctuations of the drag are ignored. This is particularly vital for stationary organisms for which the extreme drag and not the mean drag is relevant for dislodgement or breakage. This is particularly true for wave-swept rocky shores where instantaneous flow velocities of more than 20 m/s and accelerations of more than 500 m/s² may impose extremely high loads on organisms and their habitat. Additional difficulties in physical modelling arise from the fact that such organisms are typically very small (order of cm to dm) as compared to other organisms in terrestrial or marine habitats protected from waves.

5.4.5   Sediment transport

Sediment transport is a crucial process in many environmental and engineered systems. Engineering projects involving aquatic systems often require an understanding of sediment-related issues before implementing construction or restoration, including dredging of shipping channels, construction of piers, breakwaters, and harbours, and restoration of eutrophic lakes and estuaries. Additionally, contaminants in aquatic systems are often tied to sediment transport. Therefore, sufficient knowledge of sediment transport dynamics is required before the fate of contaminants can be reliably estimated. The particle terminal fall velocity or settling velocity in quiescent water is one of the most important parameters in the sediment transport equation. Its value is determined by size and density of the particle. But particle size distributions are non-uniform and for cohesive sediments flocculation makes the system more complicated. In general, particle sizes and, in the case of cohesive sediment, densities vary in space and time. This is an important source of uncertainty for models results.

Numerical simulations are often the most efficient and practical methods for predicting sediment transport in complex hydrodynamic systems. Numerical modelling requires descriptions of the individual processes involved in sediment transport, i.e. erosion, deposition and movement of sediments in the water column. Erosion, Er (kg/m²s), is the flux of particles from the sediment bed into the overlying water column, while deposition, De (kg/m²s), of particles back to the sediment bed. The movement of sediment particles in the water column is due to convection, turbulent diffusion and settling. An accurate treatment of the interaction between of these processes allows improved predictions of sediment transport in aquatic systems (Toorman *et al.*, 2007).

Measurements of sediment concentration are insufficient to determine erosion and deposition fluxes which are essential for predicting both sediment and contaminant transport. To predict erosion and deposition accurately, these quantities must be determined as functions of the sediment characteristics and hydrodynamic variables by using experiments. As bottom sediments erode, a fraction of the sediments are suspended into the overlying water and travel as suspended load, while the remaining sediments saltate in a thin layer near the bed as bedload. The fraction in each mode of transport depends on particle size and local hydrodynamics.

For fine-grained particles (which are generally cohesive), erosion occurs both as individual and aggregates of particles. Individual particles may move as suspended load, but aggregates may initially move as bedload while they disintegrate into small particles in the high stress boundary layer near the bed. It is assumed that fine-grained sediments less than approximately 200 μm are completely transported as suspended load. Coarser, noncohesive particles (defined here as those particles with diameters greater than about 200 μm) can be transported both as suspended load and bedload, with the fraction in each dependent on particle diameter and shear stress.

The initiation of motion and sediment transport must depend on, at least: boundary shear stress, sediment and fluid density (buoyancy), and grain-size. For particles of a particular size d, the shear stress at which suspended load is initiated is defined as τ_{cs}:

$$(Pa): \tau_{cs} = \begin{cases} \dfrac{1}{\rho_w}\left(\dfrac{4w_s}{d^*}\right)^2 & \text{for 200 μm} \leq d < \text{400 μm} \\ \dfrac{1}{\rho_w}\left(4w_s\right)^2 & \text{for } d > \text{400 μm} \end{cases} \tag{5.9}$$

where d^* is the non-dimensional particle diameter calculated from $d^* = d[(\rho_s/\rho_w -1)g/v_k^2]^{1/3}$, g (m/s²) is the acceleration due to gravity, v_k (m²/s) is the kinematic fluid viscosity, w_s is the settling velocity, ρ_s is the sediment density and ρ_w is the water density.

For $\tau > \tau_{cs}$, sediments are transported both as bedload and suspended load, with the mass fraction in suspension increasing with τ, until all sediments are transported completely as suspended load.

As the ratio of shear velocity, defined as $u^* = \sqrt{\tau_0/\rho}$ (m/s), to settling velocity increases, the proportion of suspended load to total load transport, q_s/q_{tot}, increases.

For suspended sediments, the time-dependent transport equation in the water over the bed is

$$\frac{\partial(hC_s)}{\partial t} + \frac{\partial(huC_s)}{\partial x} + \frac{\partial(hvC_s)}{\partial y} + \frac{\partial(hwC_s)}{\partial z}$$

$$= D_h\left[\frac{\partial}{\partial x}\left(h\frac{\partial C_s}{\partial x}\right) + \frac{\partial}{\partial y}\left(h\frac{\partial C_s}{\partial y}\right)\right] + D_v\frac{\partial}{\partial z}\left(h\frac{\partial C_s}{\partial z}\right) + Q_f \tag{5.10}$$

where C_s is the sediment concentration, D_h and D_v are the horizontal and vertical turbulent diffusion coefficients (m²/s), respectively, and h is the local depth (m). The net flux of sediments into suspension, Q_f (kg/m²/s), is calculated as the total erosion flux into suspended load minus the deposition flux from suspended load for each sediment size class (James et al., 2008).

5.4.6 Measurements and data interpretation

Most laboratory experiments studying the interaction between hydraulics and ecological phenomena require measurement of one or more of the following: (i) fluid flow; (ii) wave characteristics; (iii) sediment transport; and (iv) ecological parameters. Sediment transport measurements are detailed in chapter 4, wave measurements are detailed in chapter 2 for wave modelling in flumes and basins.

Fluid flow measurements need to characterise the mean flow parameters and also usually turbulence across a range of spatial and temporal scales which may require instrumentation capable of measuring at high sampling rates and in small sampling volumes. Experiments incorporating animals and plants may also feature extreme measurement conditions such as very shallow or fast-flowing water which will require suitable measurement techniques. Sampling volumes, spatial resolution and acquisition frequencies should be defined a priori, taking into account for the features (length scales and time-variability) of the phenomenon being studied.

To measure turbulence, the relevant fluctuating physical properties must be recorded in a highly-resolved, four-dimensional domain (three space and one time dimension). A full description of velocity also requires three components of the velocity to be measured (downstream, vertical and lateral). Furthermore, the measurements must be of sufficient intensity and duration to yield robust statistical properties such as the turbulent kinetic energy (TKE) or the turbulent energy dissipation into heat. Such observations are not always possible and, particularly, do not usually cover more than two dimensions. Typical instrumentation are as follows:

1   **Velocimeters;** measure velocity time-series at a fixed point such as Acoustic or Laser Doppler Velocimeters (ADV and LDV) and Hot Wire Anemometry (HWA). An ADV can usually measure 3-components of velocity in flows deeper than about 0.05 m and in certain configurations can measure 2-components of velocity in shallower water. LDV is capable of 1, 2 or 3-components velocity measurements using a variety of different optical configurations. The sampling volume for both types of Doppler Velocimeters is remote from the instrument which has the advantage of reducing the effect of the measurement device on the flow being measured. Typically, an ADV sampling volume size is 1 cm^3 whilst a LDV has a much smaller sampling volume which depends on the optical configuration. HWA differs significantly from LDA and ADV since it is more intrusive, but it may be particularly suitable for measurements within vegetation since the remote location of ADV and LDA probes means that acoustic or laser paths may be obstructed by moving vegetation making measurements difficult. HWA are available in a number of different configurations enabling 1, 2 or 3 component velocity measurements.

2   **Current profilers;** measure flow velocity in multiple layers throughout the water column, with ranges from a few dm to several hundred meters. The most common type of profilers use acoustic measurement techniques such as Acoustic Doppler Current Profilers which also have the capability to measure the bed elevation simultaneously with velocity measurements. Different configurations can be used to measure 1, 2 or 3 components of the velocity at each height.

3   **Particle Image Velocimetry (PIV);** measures flow velocity in a two-dimensional plane by illuminating the flow using a laser light sheet and taking repeated images using video cameras to measure the displacement of fluid tracers from which the flow velocity can be calculated. Using a single camera, two components of

velocity can be quantified in the measurement plane and two cameras may be used to measure three velocity components. The sampling volume depends on the light sheet thickness, the camera resolution and the size of the image area. PIV can also be extended to use Laser Induced Fluorescence (LIF) which is an optical measurement tool to acquire whole-field concentration data. Quantitative analysis is based on calibration images that map the response at every pixel of the camera with respect to both local concentration and local laser light energy. Quantitative information can be obtained in many concentration applications, including for example the mixing performance of e.g. chemical processes, the role of coherent mixing structures on the transport of a marker, the interaction between large-scale and smaller-scale in turbulent mixing, the dispersion of pollutants in model ocean systems, etc.

Method 1 covers one-dimensional space (time), method 2 accommodates two dimensions (space and time) and method 3 represents three-dimensions (two-dimensional space and time). All methods miss at least one spatial dimension. Turbulence is highly intermittent, so that spatial or temporal under sampling is always an issue. Furthermore, none of the methods above are able to cover the full range of spatial scales from the smallest dissipative eddies (typically of the order of a centimetre) to the largest energy-containing eddies (typically of the order of several meters). Thus, to quantify statistics of turbulence, additional assumptions have to be made. The extrapolation to the four dimensional space usually assumes local isotropy (that is, the turbulence has no preferred direction), at least on the smallest scales. To overcome the problem of only covering a part of the turbulence spectrum, it is assumed that larger scales do not significantly contribute to the micro-structure turbulence and that the smaller scales do not contribute to the energy-containing scales (Burchard et al., 2008). Biological and physical processes and their interactions over a broad range of scales, from micro-layers to ocean basins, can be evaluated. High-resolution observation methodologies, as well as applications of different mathematical tools for analysis and simulation of spatial structures, time variability of physical and biological processes, and individual organism behaviour, should also be considered.

5.5 INCORPORATING PLANTS INTO PHYSICAL MODELS

Aquatic plants are common in rivers and streams as well as in estuaries and coastal zones and the interactions between fluid flow and plants are very complex, being dependent on properties of the flow (depth, velocity, turbulence and water temperature), properties of the plants (biomechanical properties, growth form, biomass and spatial distribution), and the boundary conditions (morphology and boundary roughness). There is a wide literature describing the interaction between plants and flowing water with contributions from both engineers and ecologists (e.g. Sand-Jensen et al., 1989; Sand-Jensen, 1998; Nepf, 1999; Ghisalberti & Nepf, 2002; Ghisalberti & Nepf, 2006; Naden et al., 2006; Gurnell et al., 2006; Nepf & Ghisalberti, 2008). These studies involve either field measurements or flume experiments. The main objectives driving this research are from a hydraulic engineering perspective:

- Identify and quantify the flow processes and mechanisms that influence and control vegetation-induced stresses (i.e. flow resistance).
- Develop methods for calculating vegetation-induced flow resistance caused by various types of vegetation (flexible and rigid, emergent and submerged, over hanging bank-side). In fluvial environments this applies to vegetation in channels and on banks as well as on floodplains. In coastal environments submergence depends on water depths and tidal effects.
- Determine the affect of vegetation on wave attenuation in coastal environments.
- Develop turbulent model equations and determine drag and turbulent model coefficients to include in numerical models.
- Identify and evaluate the role of plants in sediment flux/retention and associated accretion processes.
- Determine the vegetation influence on mass flux, mixing processes and water quality.
- Assess the applicability and performance of flume based methods and equations to the natural environment.
- Test and modify traditional conveyance estimation methods for applicability and reliability or derive new conveyance estimation methods for vegetated rivers.
- Develop criteria and procedures for river design, restoration and maintenance for flood control and habitat improvement.

and from an ecological perspective:

- Determine the ranges of ambient flow and/or wave conditions aquatic plants can tolerate.
- Determine whether trade-offs exist between the structural strength of plants and their ability to reduce drag.
- Quantify the effect of hydrodynamic forces on the morphology (size/shape) of plants and potential for adaptation.
- Explore evidence for parallel evolution of form (plant shape) suited to drag reduction across aquatic macrophytes.
- Determine breaking strength of plant stems.
- Determine conditions for uprooting plants.
- Determine transport and dispersion of seeds and propagules.
- Determine plant patterns and distribution in different flow and/or wave regimes.
- Assess why field measurements and laboratory experiments on the influence of water velocity on photosynthesis/primary production show conflicting results.

5.5.1   Challenges of physical modelling using plants

The main challenge relating to experimental design is the reproduction of plant and flow properties which are representative of field conditions. This section discusses the key variables describing plants and the options available for their simulation in physical models. It is important to recognise that plants are highly variable both between and within species – they grow and change on a seasonal timescale both in terms of biomass and biomechanical properties; they alter their shape and form with flow conditions; and the plants present in any particular environment may vary from year to year dependent on hydrological conditions and water quality. The plethora of different attributes also means that flume experiments need to be well designed to focus on key plant-flow relationships whilst being practical regarding the number of experiments which it is feasible to undertake. It is also essential to recognise that plant attributes may be described in different terms by different disciplines.

Plant types

Notwithstanding the number of individual aquatic plant species, these can be classified into a relatively small range of common forms or morphotypes. Those recognised by the River Habitat Survey in the UK (Environment Agency, 2003) are given in Table 5.2 and some common forms are shown schematically in Figure 5.3 which illustrates their occupancy of the water column and their spatial distribution vis-à-vis the channel. This classification is also appropriate for estuaries and other shallow water environments.

Emergent plants may vary considerably in the frequency and depth of their inundation. On floodplains for example inundation may be annual and shallow, whereas within channels it is almost continuous and of variable depth and velocity. Floodplain vegetation may be tall (e.g. trees), dense (e.g. bushes) and heterogeneous

Table 5.2 River Habitat Survey in-stream morphotypes.

Liverworts/mosses/lichens	Emergent broad-leaved herbs
Emergent reeds/sedges/rushes/grasses/ horsetails	Floating-leaved (rooted)
Free-floating	Amphibious
Submerged broad-leaved	Submerged linear-leaved

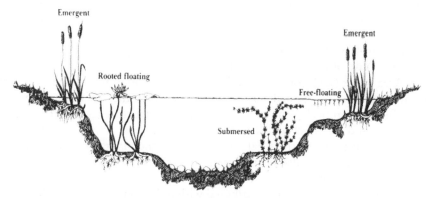

Four groups of aquatic flowering plants

Figure 5.3 Four groups of commonly occurring aquatic flowering plants (after Managing Iowa Fisheries, 2009).

which adds further complications to the predictions of flow resistance. Numerically, the simplest approach may be to model vegetation as rigid vertical cylinders (e.g. Nepf, 1999). However, whilst this may be a reasonable model for rigid trees, assuming that vegetation behaves as a rigid cylinder will result in significant errors in the relationship between flow velocity and drag force if the vegetation is flexible (e.g. Fathi-Maghadam & Kouwen, 1997, see also section 5.4). The degree of flexibility may be affected by the density of the vegetation and, on flood plains, the flexibility of plants may vary with inundation depth. This will affect the overall flow resistance. For example, flume experiments have shown that real, flexible vegetation shows increasing flow resistance as flow depth approaches vegetation depth and tends towards a constant value at submergence depths greater than twice the vegetation depth (e.g. Wilson, & Horritt, 2002).

Plant attributes

Experiments are generally designed to focus on individual morphotypes rather than simulate the complexity of the field case. Focussing on single morphotypes may be appropriate, since recent fieldwork based on UK rivers (Gurnell *et al.*, 2010) has shown that the occurrence of different vegetation morphotypes (mosses, linear-submerged, patch-submerged, linear emergent, branched emergent) can be discriminated by stream power and sediment size. For any given morphotypes we need to define several different attributes:

Stem spacing (often referred to as plant density or porosity): This affects whether stems interact with the water as individuals or as a group. For example, at low stem spacing, a drag term will relate to that of a single stem, whereas at high density, the effect of individual stems of the flow will interact and the drag terms needs to be defined in relation to a bulk drag coefficient. This issue can also be extended to the question of whether experiments are focused on individual plant stems, a field of plant stems in either regular or irregular configurations or, in the case of large plants, whole plants.

Biomass is the total mass of living matter within a given bed area and is a standard biological measurement describing temporal/seasonal changes in plant growth. This attribute may be particularly useful for experiments designed to examine the effects of vegetation on bulk flow properties such as the effect of vegetation management strategies (cutting or removal) on discharge capacity (e.g. Vereecken *et al.*, 2006). Similarly, biomass could be a useful measure for experiments investigating the ecological response of plants to a given flow regime.

Leaf Area Index (LAI) (or leaf area density) is the ratio of total upper leaf surface of vegetation divided by the bed surface area over which the vegetation grows. This can be useful for quantifying the influence of canopy structure (constant versus variable leaf areas) and the effect of differences between species, plant age and/or seasonality. Expressed in terms of the total wetted area of the plants per unit bed area or water volume, the LAI can also be an important resistance parameter.

Morphology of plants describes the dimensions of stems and leaves (e.g. strap-like or filamentous) and how they are distributed within the water column under different flow conditions. This includes linear characteristics (length, width and deflected canopy height); areal characteristics (front-area, side-area, plane-area and cross-sectional area) and volumetric characteristics (plant volume per unit water volume).

Stiffness, flexibility and reconfiguration of stems and leaves will vary with flow conditions, seasonality or age of the plant, its nutrition and light availability. They will be represented by biomechanical properties of the plants such as plant material density and Young's elasticity modulus. As well as the propensity for plant stems to bend, attention must also be given to the movement of the plant in the flow and whether this involves movement of the whole plant or just the leaf ends. For example, whether the plant is simply responding to the flow or whether there is transmission of momentum between the plant and the flow. Seasonal plant growth will change the volume, stiffness and total biomass (Neumeier, 2005) and must therefore be taken into account.

Patchiness or spatial distribution of plants in the field is another critical consideration. Experiments have often been conducted with uniformly-spaced fields of grasses or even plastic surrogates. Relatively little detailed hydraulic work has been carried out on larger leafy plants which form organised hydraulic environments often resembling flow patterns more generally found in a braided channel. The position of plants within the flow will also have an effect e.g. in rivers, marginal locations versus mid-channel. Combinations of different plant forms (as depicted in Figure 5.3) have not been studied in experiments and the interaction between different ecological components is an important area for further research.

Biological measurements

The plant attributes described above involve a range of biological measurements. Some examples of measurement techniques from recent literature are summarised in Table 5.3.

Table 5.3 Examples of biological measurements to describe plant attributes.

Leaf length	Measure with a ruler (Gacia et al., 1999).
Leaf area index	Calculate ratio of leaf area to ground area (Gacia et al., 1999).
Shoot and stem densities	Randomly place a quadrat of a known area and count the number of shoots/stems in the total area (Leonard et al., 1995; Gacia et al., 1999). Alternatively, all the plants within the quadrats can be removed and the shoots and stems separated and counted back in the laboratory to calculate shoot density (number of shoots/area of quadrat) or stem density (number of stems/area of quadrat) (Neumeier, 2005). These measurements may also be taken at different vertical positions to obtain a vertical profile (e.g. Lee et al., 2004)
Total biomass	Measure dry weight by oven drying for 24 hours at 60°C (Gacia et al., 1999) or for 48 hours at 80°C (Neumeier, 2005).
Shoot, stem and thallus diameters	Measure with callipers or plastic measuring tape (e.g. D'Amours & Scheibling, 2007).
Vertical biomass distribution	Useful for understanding lateral obstruction of plants (Neumeier, 2007). Calculate the dry weight for vertical sections of the plants which equates to weights per layer per ground unit. This can be described as the percentage of total biomass per layer and can be used to measure relative distribution of the canopy. (Neumeier, 2005, 2007).

(Continued)

Table 5.3 (Continued)

Lateral obstruction Vegetation denseness Bushiness index	A rigid coloured plastic background is inserted vertically, parallel to the vegetation being measured. A mirror equivalent to the canopy height being measured is inserted vertically at an equivalent distance and tilted to 45° using a triangular wooden frame. A photo of the reflection is then taken from above and the vegetation is differentiated from the background using image analysis software. A binary image is produced with presence or absence of vegetation allowing total lateral obstruction/vegetation density/bushiness to be calculated as obstructed area/picture width (Neumeier, 2005). A bushiness index can also be calculated using the ratio of plant circumference-to-length (D'Amours & Scheibling, 2007).
Forces exerted on plants Plant movements	Force transducers placed at the base of plants can be used to record the forces exerted on plants (e.g. Stewart, 2006). A Video camera can be used to record plant motion which in conjunction with image analysis software may be used to calculate the velocities of plant movements (e.g. Stewart, 2006).

5.5.2   Best practice guidelines for plants in flume experiments

Having identified the questions to be addressed in the experiments and the plant attributes and setting of interest, we now need to select the best means of achieving good experimental results. Key considerations are the scale of the experimental facility with respect to the plants and the choice of plant.

Scale

The size of experimental channel available will limit the type of experiment and choice of plant or surrogate. Nowell & Jumars (1987) present a good overview of the key theoretical and experimental considerations required for simulating benthic environments. A key consideration is the length of the flume which needs to be sufficiently long to: (i) avoid entrance effects, which may persist for at least 20 times the critical length scale (flow depth or inlet pipe diameter) before a well-developed boundary layer flow is established; (ii) avoid backwater effects from the outlet; and (iii) leave a reasonable length of channel for experimentation. The flume also needs to be sufficiently wide to avoid significant sidewall effects. A number of authors have demonstrated the effects of width on experimental results (e.g. Naot & Rodi; 1982; Nowell & Jumars, 1987; Johnsson *et al.*, 2006). A minimum requirement is that measurements are taken well away from wall boundary layers but ideally the width should be greater than 5 times the depth to minimise wall effects. The flow depth should be large enough to ensure the required distribution of the plant within the water column (e.g. emergent and submerged cases) but care should be taken to maintain an appropriate aspect (i.e. width-depth) ratio. The need for hydraulic scaling in terms of Reynolds number and Froude number have been addressed in section 5.4.

Choice of plant or surrogate

There are three options available when selecting the type of vegetation to use in an experiment: (i) some type of artificial or surrogate plant; (ii) scaled plants or the use of natural plants. Some examples of their use from the literature are given in Table 5.4. The choice will have to bear in mind the scale of the flume and the purpose of the experiment.

The advantage of using artificial or surrogate plants is that they are easy to use, inert and can be manufactured to a precise specification. Their use also tends to focus on individual plant attributes e.g. rods or wooden dowels for stem density or plastic strips or straws for flexibility and reconfiguration. Rigid cylinders can be used to represent stiff vegetation and can be scaled on the basis of their diameter and different spatial arrangements can easily be reproduced (e.g. McBride *et al.*, 2007). Their disadvantage is the question of whether they are sufficiently similar to aquatic vegetation to be applicable to the field case. Clearly they will be inappropriate for some ecological experiments. Where plants have some flexibility it is essential to ensure that the biomechanical properties of the plant are reproduced in the surrogate that is chosen. This may require careful selection of materials and may involve some compromise which reinforces the need to understand the objectives of the study to ensure that the appropriate characteristics are reproduced in the surrogate that is used. Several studies have used strips of different plastic materials (e.g. PVC) to represent blade-like vegetation (e.g. Sand-Jensen, 2003; Folkard, 2005). Folkard (2005) chose a polyethylene sheeting (Decco) as a surrogate that closely replicated the density and modulus of elasticity of natural seagrass plants. A more complex approach to producing surrogates was used by Stewart (2006) to model reef algae. This involved taking casts of typical reef algae using plaster of Paris and creating models using a two part epoxy. These models were then used to make reusable moulds from silicone which were

Table 5.4 Choice of plant.

Choice of plant	Purpose
Artificial: rods or wooden dowels	Stem density effects on drag and flow resistance (e.g. Nepf, 1999, Stone & Shen; 2002; James et al., 2004). Flow resistance on flood plains (e.g. Pasche & Rouvé, 1985; Naot et al. 1996; Rameshwaran & Shiono, 2007).
Artificial: rods with strips	Flow structures (e.g. Nepf & Ghisalberti, 2008; Nepf & Vivoni, 2000).
Artificial: plastic strips or strips with foliage	Vegetation-flow interaction (e.g. Wilson et al., 2005).
Artificial: plastic bushes or grasses	Floodplain roughness on flow structures, bedforms and sediment transport rates (e.g. Shiono et al., 2009a, b).
Scaled plants (e.g. aquarium or nursery grown plants)	Flow resistance (e.g. Järvelä, 2002).
Natural large aquatic plants (e.g. Ranunculus plant)	Plant-flow interaction (e.g. Naden et al., 2004).
Natural vegetation (e.g. grass)	Flow resistance (e.g. Stephan & Gutknecht, 2002; Wilson & Horritt, 2002, James et al., 2004, Carollo, et al., 2005).

then used to construct models from different materials to create algae with different buoyancies and flexural stiffness attributes (very flexible, flexible, flexible non-buoyant, extra buoyant stiff and rigid).

Using natural plants on the other hand avoids the issue of ensuring comparable properties but natural plants are highly variable and this may make it necessary to carry out replicate experiments. Additional issues that arise are the fixing of the plants within the flume and the need to maintain their state of health during the experiments. It also needs to be remembered that their availability, biomass and biomechanical properties will depend on the season and possibly on climatic conditions, hydrological regime, water quality and any management that may have taken place. It will also be necessary to obtain permission from the site owner and the appropriate regulatory agencies (e.g. Natural England and the Environment Agency in England). Dependent on the plant of interest, the flume dimensions will also be a limiting factor and a potential solution is to use scaled plants, e.g. aquarium species, with similar attributes to the desired prototype.

Finally, dependent on the purpose of the experiment, a choice between single stems or whole plants needs to be made.

Plants in the flume

As described above, flume experiments may either be conducted with real plants or with artificial plant surrogates. The substrate of the flume will need to be chosen to be appropriate to the experiment e.g. smooth bed, fine-grained roughness or gravel. The plants or plant surrogates need to be fixed relative to the substrate in a similar manner to that found in the natural environment. Artificial plants may be fixed to PVC plates (Neumeier, 2007) or to Plexiglas plates using silicone sealant (Graham & Manning, 2007). In case of natural plants, it is essential to keep them in a good, healthy condition as found in the natural environment without losing biomass or deterioration in their biomechanical properties over the duration of experiment. For short duration experiments, an artificial fixing may be sufficient to meet this requirement e.g. using chicken wire embedded in gravels. An alternative and certainly a requirement for longer-term experiments is to pot-up the plants into a suitable material and bury the pots within the flume substrate.

There is also the potential to grow plants in the flume (e.g. periphyton) with appropriate control on bed material, temperature, light and nutrients (e.g. Bouma et al., 2005). However, clean flume water or substrate may be detrimental to plants. In all cases, using natural plants in flume experiments needs proper attention to simulating the natural environment of the plant (including water chemistry, water quality, nutrients, light and temperature) and replicating experiments in cases where variability cannot be controlled. This is a particular problem for marine and estuarine plants as few laboratory flumes are designed for saline water.

Flume set-up

Flume experiments need to be performed under steady and uniform gravity-driven flow conditions (i.e. sloping flume). The flow field measurement sections need to be located in the fully developed boundary layer region away from the inlet and outlet sections. In order to establish uniform flow conditions under a known discharge, the flume set-up needs to have a discharge measurement facility with a depth control mechanism at the outlet.

Hydraulic measurements

Flow field measurements can be made using Acoustic Doppler Velocimetry (ADV), Laser Doppler Anemometry (LDA) or Particle Image Velocimetry (PIV) around plants. Within plants, Hot-Wire anemometry (HWA) can be used (see Section 5.4.6). It is not currently possible to use PIV techniques within or too close to plants because of the masking effect of the plant. There are several measurement techniques for determining the channel boundary roughness properties of a gravel-bed surface (Smart *et al.*, 2004). Measurement techniques such as laser scanning give sufficient detail for calculating the physical properties of the bed – expressed either as a roughness height value (k_s) or as a near-bed porosity and average frontal projected and surface areas of boundary roughness elements for numerical modelling. Free surface variations can be measured using photogrammetry techniques (Chandler *et al.*, 2008). Plant morphology characteristics and motion (deflected height and area and volumetric properties) during experiments can be measured using underwater cameras. Plant attributes and biomechanical characteristics including plant-flow interaction characteristics such as drag and lift coefficients, strain, tension and bending moments (Nikora, 2009) need to be measured or quantified during experiments using appropriate measurement techniques.

5.5.3 Developing best practice

The use of plants in flume experiments is diverse in terms of the types of hydraulic conditions that may be modelled and the plants that may be used. Nevertheless, some basic guiding principles can be proposed:

- Ecological and (or) biological expertise relating to the plants of interest should be sought in order to maximise experimental efficiency. Expert guidance will be valuable in relation to many of the issues listed below;
- Where natural plants are used, experimental conditions may need to be carefully controlled so that plants behave as they would in the natural environment. This is necessary to ensure the validity of the experimental results and requires consideration of both environmental conditions (e.g. temperature, light, water quality and chemistry) and ecological conditions (e.g. appropriate plant densities and arrangements);
- Where surrogates are used, careful consideration needs to be given to the full set of biomechanical characteristics of the surrogate material which needs to be representative of the natural counterpart;
- Wherever possible, experimental work should be supplemented by field research so that experimental studies represent the key characteristics of the natural environment;
- Scaling issues need to be carefully considered and may require the use of surrogates;

- Select species that are representative of plant groups, if at all possible, in order to maximise the potential transferability of results;
- Include the necessary hydraulic expertise so that experiments are conducted in strict accordance with hydraulic theory in order to ensure that results reflect the processes rather than an inappropriate flume set-up.

5.6   INCORPORATING SMALL ANIMALS INTO PHYSICAL MODELS

Increasingly, society requires sustainable management of aquatic resources wherein engineering interventions are evaluated not only by their engineering effectiveness but also by the degree to which they work in harmony with natural systems, accommodating, for example, the habitat requirements of fauna and flora. Interdisciplinary research, particularly the development of methodologies that can yield sustainable solutions, is therefore important. Empirical field studies, numerical modelling of various kinds and physical modelling in experimental facilities can be used in complementary and reinforcing ways, to develop better understanding of the interactions between animals and their hydraulic environments.

Within this broad context, this section considers the incorporation of small aquatic animals into hydraulic studies in experimental flumes and tanks. We define small animals as sessile and labile benthic invertebrates with principle dimensions on the order of 10^0 to 10^2 mm (e.g. some crustaceans, molluscs, insect larvae, echinoderms). These animals are ubiquitous in the hydrosphere, playing a fundamental role in the ecology of marine and freshwater ecosystems. Many of them are also important agents of physical and chemical change within their ecosystems, moving sediment, affecting flows and modifying water quality. In ecology, there is a well-established body of work (stretching back at least one century) which considers the ways in which animals are affected by and interact with the hydraulic aspects of their environment (see Vogel, 1994). Relatively less work has examined how animals affect the hydraulics of the environments where they live, but just as plants are recognised as important roughness elements, sources of coherent flow structures and sinks for fine sediment, so small animals may be important in order to understand flow patterns and processes, especially at small scales. In some cases, as with aquatic plant experiments, physical analogues may be used to represent animals, but in other cases the behaviour of organisms may be important, necessitating the incorporation of live animals in physical experiments.

5.6.1   Challenges of physical modelling with small animals

Experiments involving small animals present the same challenges that are familiar in purely physical modelling experiments (scaling up results, technical difficulties, realism) and some new ones including the importance of biological processes, identifying hydraulically meaningful biological groups, accommodating the singular behaviour of individuals and species, understanding biological as well as physical stimuli, and the husbandry and welfare of live organisms. Several of these issues are considered here. See Rice *et al.* (2010) for more detailed consideration of the challenges for experimentalists working at the interface of river ecology and hydraulic engineering and Jonnson *et al.* (2006) for a marine science perspective.

Realism and simplification

The primary strength of physical experimentation is the ability to control some variables and manipulate others. This strength is also a significant drawback, however, as simplifications are always at the cost of realism. One particularly exciting aspect of work involving small animals is the opportunity to independently manipulate physical and biological parameters to understand their relative influence on system processes: for example, manipulating the density of organisms to measure the relative influence of biotic and mechanistic processes on boundary hydraulics or sediment entrainment stresses. However, a substantial problem when working with live animals is that animals in flumes may behave unnaturally in response to the simplified flume environment and to invasive measurement instruments. The former may include simplification of the physico-chemical environment or the biotic environment (food resources, competitors, predators, etc.).

A few studies have examined the response of freshwater and marine benthic invertebrates to the conditions in artificial tanks and flumes (Barmuta *et al.*, 2001). For example, in an examination of six species of marine molluscs, crustaceans and echinoderms, Ofstad (2002) found that only two would feed on food which was different from their natural food and that two species did not feed at all. Notwithstanding observations like these, there is a dearth of systematic information about the impact of experimental conditions on animal behaviour. In addition to dedicated experimental programmes designed to fill this gap, best practice should be to routinely include such evaluations in any experimental programme, even though it may be impossible to eliminate such biases completely or derive a correction factor of any sort. The key challenge here is to minimise the stresses that are placed on animals by their artificial environment, to try and identify any modifications of behaviour and ensure that these do not compromise the study objectives, interpretations and inferences.

In turn, useful data will depend on adequate consideration of levels of both biological and physical abstraction and, while engineers will be familiar with issues of physical realism, best practice should be to consult with appropriate ecological expertise in order to understand what may be lesser known environmental controls on the behaviour of small animals. For example, most flume facilities are brightly lit, but many invertebrates are sensitive to light conditions with distinctive activity regimes at different times of day, so it is important to manage lighting carefully.

Scale effects

Scale is an important element of simplification and realism. To ensure appropriate hydraulic conditions some attempts have been made to define appropriate flume dimensions when working with small animals and fish; in part to guide ecological experiments toward flume dimensions that produce realistic boundary layers (e.g. Nowell & Jumars, 1987). In an important recent review, Jonsson *et al.* (2006) consider explicitly flume requirements for biological experiments and recommend that flume width should exceed $2\delta + k_l$, where δ is the boundary layer thickness and k_l is a relevant length scale of the organism under investigation. Satisfying such recommendations should be a minimum requirement that ensures sensible hydraulic conditions.

For many small animals, experiments in flumes and tanks can be at or close to prototype scale (that is, the physical environment is not a scaled version of reality). However, the absolute dimensions and facility for introducing realistic spatial complexity (patchiness) often are, in all but the very largest facilities, constrained by the size of the flume or tank. This may be problematic because both aspects of the spatial domain are important for numerous ecological processes and functions, such that constraint may modify behaviour. Again, it is prudent to take expert advice on this in order to minimise adverse impacts on experimental results.

More generally, if experiments involving small animals are incorporated into physical models that are constructed at less than prototype scale, there is the potential for geometric, kinematic or dynamic scale effects to influence experimental observations. The traditional use of similarity principles to minimise these effects, can be extended to animal facsimiles if live animals are not required and it is sufficient to represent animal morphology using a scaled model, but scaling principles do not extend to live animals. The challenges of ensuring consistency between model and prototype that are apparent in purely physical experiments (e.g. Peakall et al., 1996) and that are substantially increased in experiments that involve living plants (cf. Tal & Paola, 2007), are amplified further by the inclusion of living animals. Sensible scale similarities may be intractable because identifying a scale version of the live organism that retains equivalent morphological and behavioural characteristics is impossible. Moreover manipulation of fluid properties such as water temperature to minimise viscous effects and achieve sensible turbulent conditions, is likely to have inadvertent effects on animals. This does not imply that all experiments must be at prototype scale because scaling for all hydrodynamic properties may, in any case, be impossible and, moreover, much can be learned from heavily distorted or entirely un-scaled experiments. Jonsson et al. (2006) provide an extended discussion of scaling flow properties in experiments involving animals.

Species diversity and generality

It is difficult to generalise the results of a study using one species to other species. Even species that are closely related taxonomically may exhibit substantial differences in physiology, morphology and behaviour that influence their interactions with hydraulic environments. The example of fish passes, where research biases toward salmonid fish have negative consequences for other species, illustrates that there may be real "costs" involved (Rice et al., 2010). Attempts have been made to improve generality by classifying organisms according to their morphological adaptations for life in flowing water, or hydraulic traits, but rigorous empirical studies linking traits to hydraulics are rare. Even within groups that are assumed to interact with hydraulic phenomena in a consistent manner, the behaviour of individuals cannot be ignored. Experimental replication with different individuals or groups of individuals is therefore important, not only to satisfy assumptions of statistical testing but also because animals are capable of learning and modifying their behaviour.

Flume facilities and hydraulic measurements

Jonsson *et al.* (2006) compared several experimental flume and tank facilities to determine their relative strengths and weaknesses for studying benthos-flow interactions in marine environments. They examined straight, annular, 'race-track' and field flumes in 13 institutes that formed the EU network, BioFlow and found that the flumes varied in their ability to develop realistic boundary layers and turbulence characteristics. Beyond this work we are unaware of general attempts to define facility protocols, flume designs or best-practice procedures where the intention is to use live animals. Through-flow flumes that utilise natural river water are particularly useful for long-term studies of animal behaviour and performance because water quality is relatively easy to maintain, but the potential to attain high test velocities is likely to be limited. In contrast, maintenance of water quality and temperature (pumps may heat the water) in traditional wave tanks and re-circulating flumes, can be problematic, requiring water cooling machinery.

Making robust hydraulic measurements close to small animals is challenging, especially where it is necessary to acquire spatially distributed information at relatively high resolution. The practical challenge of collecting hydraulic measurements at scales no greater than the length scale of the animals involved is not necessarily a difficulty. Indeed, impressive detail is possible using, for example, hot-wire anemometry. However, where the focus is not on hydraulics in the immediate vicinity of an animal but on a hydraulic domain that is many times, perhaps orders of magnitude, larger than the animal, this challenge grows. Point sampling on large grids can yield useful information but is time-consuming and therefore expensive. Particle Image Velocimetry (PIV) provides high-resolution information relatively more rapidly but seeding materials and laser parameters may have adverse behavioural effects, cause tissue damage or death. Stamhuis *et al.* (2008) discusses many of the practical problems and benefits of using PIV with live animals. An alternative to extensive direct measurements is to utilise numerical modelling, specifically Computational Fluid Dynamics (CFD) calibrated against coarsely distributed measurements, but this also remains challenging; for example where boundary roughness is complex.

Animal welfare and ethical considerations

Taking care of animal welfare before, during and after their use in experiments is both an ethical obligation and necessary to minimise stress and thereby promote natural behaviour. It is difficult to recognise signs of stress in small animals like molluscs and insect larvae, even though there is no logical reason to suppose that their nervous systems are any less complex or their sensory capabilities any less sensitive than in larger animals. Perhaps because they are so distantly related to humans (unlike, for example apes) there may be an assumption that stress and pain are not 'felt' by such animals. It is important to avoid these unfounded, value judgements and recognise that if we cannot recognise stress, it does not follow that animals are not stressed. Best practice should focus instead on giving all animals appropriate care and attention, minimising handling and carefully tuning holding and experimental conditions to maximise their well-being.

This requires a basic understanding of the animal's biology, what conditions they might find stressful and what their basic needs and tolerances are. At the very least, require oxygen, food (energy) and an appropriate physical environment (e.g. sandy, rocky or vegetated substrate). Expert advice should be sought to establish details for particular species, although some generalities can be made. For example, if animals are kept in a tank system for some period of time, it is important to consider the oxygen demand of the animals in question and thence oxygenation requirements. Metabolic rates including respiration rate, are scaled allometrically with animal size according to $B = B_0 \cdot M_b^X$ where M_b is body mass and the coefficient, B_0 varies between taxonomic or functional groups but the scaling exponent, X is always approximately 0.75.

In many countries, the use of animals is governed by legislation and researchers are required to be accredited for experimental work. Licenses may be required to work with particular species, although this is less likely when working with invertebrates than fish, unless the species is rare or an invasive alien.

Other practical considerations

In addition, there are practical requirements that working with live animals bring to laboratory experiments which engineering facilities may not be designed to accommodate. Two of the most important are: controlling water temperature, water quality and lighting conditions, as well as flow, within the experimental facility; and keeping track of animals that are difficult to see because of size, low illumination or speed of movement. Practical solutions to many of these issues and general guidance can be found in the interdisciplinary literature.

Growing organisms *in situ* can overcome some of the issues associated with transfer of organisms. However, rearing in the laboratory also has limitations. Usually specially made tanks need to be built for this purpose (Davis, 1970). Some fish species are naturally more resilient to artificial spawning and culturing; however in all cases specific niche conditions need to be replicated in the laboratory to ensure survival. Galbraith *et al.* (2006) attempted to rear fish and found a reduction in gamete viability associated with premature activation from early contact with water. Working with juveniles for example in sediment toxicity tests, can also be problematic as this life stage is most sensitive to changes in the environment (Ducrot *et al.*, 2006). The authors looked at rearing molluscs in the laboratory and found that an acclimatisation period to laboratory conditions is required. Furthermore, the use of natural sediment can reduce behavioural and dietary stresses.

5.6.2   Developing best practice

This is a relatively novel field and best practice guidelines require development through dedicated research. Moreover, guidelines can only be very generic in the absence of specific knowledge of the organism of interest and the research questions being asked. Nevertheless, some basic guiding principles can be proposed.

- Ecological and (or) biological expertise of the animals(s) of interest should be sought in order to maximise experimental efficiency. Expert guidance will be valuable in relation to many of the issues listed below.

- Cues for animal behaviour need to be reproduced or controlled for, in order to isolate responses to the chosen experimental treatments and develop confidence in the interpretation of experimental observations. This includes both environmental conditions (e.g. temperature, light, substrate) and biological conditions (e.g. appropriate animal densities and arrangements).
- Wherever possible, experimental work should be supplemented by fundamental empirical research on organism behaviour and tolerances in natural environments so that flume effects can be identified and correction schemes developed.
- Scaling of the hydraulic parameters that are relevant to the animals involved in an experiment requires careful consideration.
- Handling and environmental stresses must be minimised, before, during and after experiments for both ethical reasons and to ensure that sensible data is collected.
- It is important to recognise that flume and tank dimensions and materials, as well as measurement techniques, may affect animal behaviour so that developing the correct flow conditions is not the only consideration in setting up an experiment.
- Select the species that are used to be representative of animal groups, if at all possible, in order to maximise the potential transferability of results.
- Replicate experiments to incorporate variability due to individuals and avoid bias associated with learned behaviours.

5.7 LINKING PHYSICAL MODELS OF ECOLOGY WITH NUMERICAL MODELS AND THEIR EXTENSION TO FIELD DATA

In aquatic ecosystems such as rivers and streams, flow processes occur over multiple scales ranging from a single aquatic element to aquatic communities. The functions of aquatic communities are mainly controlled by an interaction of biological, physical and chemical processes within the system (Nikora, 2009). In order to understand the flow processes over this wide range of scales and flow conditions and to determine the ecologically favourable flow regimes, it is essential to consider very detailed flow-field behaviour as well as the habitat characterization of such ecosystems. In practice, due to constraints on detailed sampling, ecological studies based either on physical models or field experiments can only consider a limited set of flow variables in relatively small parts of a system under limited flow conditions. Numerical modelling may help to fill this gap – either in flow fields or in up-scaling from the flume to the natural environment – provided that it is soundly based and sufficiently well validated. For example, three-dimensional Computational Fluid Dynamics (CFD) models (i.e. numerical models) may provide modelling tools to simulate flow field behaviour such as transport processes and physical interactions at a scale relevant to organisms in the aquatic environment at a variety of flow conditions for scales appropriate to the ecosystem being considered. The ability of CFD models to predict accurately in aquatic ecosystems depends on the use of appropriate model and turbulence closure equations and appropriate parameters. Some of these parameters can be derived from sampled data either from a physical model or in the field.

5.7.1   Combination with field measurements and numerical modelling

McIntire (1993) suggested that individual laboratory experiments conducted in isolation would contribute little to the overall understanding of ecosystems in natural streams. Instead the strength of laboratory experiments comes from them being a part of a holistic approach to investigating ecosystems which includes, conceptual and mathematical modelling as well as field observations and measurements. The model proposed by McIntire (1993) highlights the role of physical models within the overall framework of scientific investigation for understanding ecological problems (Figure 5.4). Physical modelling experiments of ecological systems are essential for testing hypotheses, but they need to run alongside the collection of field data either from field experiments or observations from natural systems. The model also high-

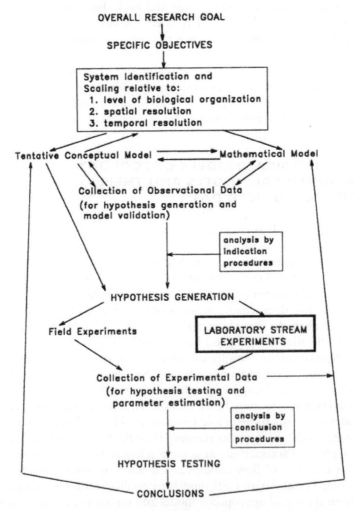

Figure 5.4 Diagram to show the integration of laboratory stream experiments with field observations and experiments and conceptual and mathematical modelling (from, McIntire, 1993).

lights the role of mathematical or numerical modelling with results from field and laboratory experiments being used to parameterise and improve mathematical models which may broaden the data available for testing hypotheses. The model should perhaps be expanded to show numerical modelling or experiments also running parallel with field and laboratory experiments to test hypotheses and improve the estimation of parameters.

5.7.2 Model equations

Traditionally, the Reynolds-averaged continuity and Navier-Stokes equations (RANS) with a suitable turbulence closure scheme are employed in CFD models to simulate flow processes (e.g. Rameshwaran & Naden, 2004a; 2004b). In the case of aquatic ecosystems such as rivers and streams, the RANS approach becomes inappropriate due to the multi-scale property of flows with complex physical and biological boundary conditions (Nikora, 2009). Double-averaging methodology potentially poses an alternative approach which has a sound theoretical basis for representing flow processes in aquatic ecosystems (Nikora et al., 2007a).

Flow through plants and over coarse bed materials may be simulated as if it were flowing through a porous medium. In this case, it is most appropriate to use the double-averaged (in time and space) continuity and Navier-Stokes equations which include drag terms, form-induced momentum fluxes, and blockage (porosity) effects resulting from the spatial averaging procedure (Pedras & de Lemos, 2001; de Lemos, 2006; Nikora et al., 2007a; 2007b). The flow variables solved by using the double-averaged equations represent flow parameters averaged in both time and space domains related to the averaged porous medium parameters which have the potential to be sampled in the field. The double-averaged continuity and Navier-Stokes equations (DANS) for turbulent flow can be written as (Nikora et al., 2007a):

Continuity:

$$\rho\frac{\partial \phi}{\partial t} + \rho\frac{\partial \phi\langle \bar{u}_i \rangle}{\partial x_i} = 0 \tag{5.11}$$

Momentum equation:

$$\phi\frac{\partial \langle \bar{u}_i \rangle}{\partial t} + \phi\langle \bar{u}_j \rangle\frac{\partial \langle \bar{u}_i \rangle}{\partial x_j} = \phi g_i - \frac{1}{\rho}\frac{\partial \phi\langle \bar{P} \rangle}{\partial x_i} + \frac{\partial}{\partial x_j}\phi\left(v\left\langle \frac{\partial \bar{u}_i}{\partial x_j} \right\rangle - \langle \overline{u_i' u_j'} \rangle - \langle \tilde{u}_i \tilde{u}_j \rangle \right)$$
$$+ \frac{1}{\rho}\frac{1}{V_0}\iint_S \bar{P}Nov_i dSwb - \frac{1}{V_0}\iint_S \left(v\frac{\partial \bar{u}_i}{\partial x_j} \right)n_j dSwb \tag{5.12}$$

where i and j are standard tensor notation indicating two out of the three x, y and z coordinate directions, $\langle \bar{u}_i \rangle$ is the time-space averaged velocity and $\tilde{u}_i$ is the spatially-varying part of the time-averaged velocity where the instantaneous velocity is decomposed as $u_i = \langle \bar{u}_i \rangle + \tilde{u}_i + u_i'$, in the x_i direction, u_i' is the fluctuating part of the velocity where the instantaneous velocity is decomposed as $u_i = \bar{u}_i + u_i'$, ρ is the density, P is the pressure, g_i is the gravity force per unit volume, ϕ is the porosity, V_0 is the total volume of the averaging domain, Swb is the extent of the water-bed interface bounded by the averaging domain, Nov is the outward (i.e. into the fluid) unit vector normal to the bed surface, the overbars denote time averaged variables and angle brackets denote the spatial averaged variable. A similar doubled-averaged transport equation for suspended sediments is given by Nikora *et al.* (2007a).

5.7.3 Parameterization of individual terms

The DANS equation (5.12) contains the following additional terms compared to the equivalent RANS equation: the dispersive or form-induced stresses $-\langle \tilde{u}_i \tilde{u}_j \rangle$ due to spatial deviations, the total drag force due to the plants and/or roughness elements which is composed of both form (pressure) drag $(1/\rho V_0)\iint_S \bar{p} Nov_i dSwb$ and viscous drag $-(1/V_0)\iint_S (v \partial \bar{u}_i / \partial x_j) Nov_j dSwb$, and porosity ϕ. For modelling purposes, individual terms in equation (5.12) are parameterized in the modelled equations.

Dispersive or form-induced stresses $-\langle \tilde{u}_i \tilde{u}_j \rangle$

It has been have shown that form-induced stresses can contribute up to 35% of the total measured shear stress dependent on the plants, roughness arrangement, physical geometry and flow conditions (e.g. Campbell *et al.*, 2005; Nikora *et al.*, 2007b; Aberle *et al.*, 2008; Manes *et al.*, 2008). However, model equations have not yet been developed to parameterize form-induced stresses (Nikora *et al.*, 2007b; Walters & Plew, 2008).

Porosity ϕ

The blockage effect on the flow by the organisms and roughness elements is accounted by the porosity ϕ which is defined as (Nikora, 2007a):

$$\phi = \phi_s \langle \phi_t \rangle \tag{5.13}$$

with $\quad \phi_s = \dfrac{V_f}{V_0} \quad$ and $\quad \phi_t = \dfrac{T_f}{T_0}$

where ϕ_s is the spatial porosity, ϕ_t is the time porosity, V_f is the fluid-only volume, V_0 is the total volume of the averaging domain, T_f is the averaging time interval equal to the sum of time periods when a spatial point under consideration is occupied by fluid only and T_0 is the total averaging time interval including periods when the spatial points are intermittently occupied by fluid and organisms.

Drag force terms $\left(Fd = (1/\rho V_0)\iint_S \bar{p} Nov_i dSwb - (1/V_0)\iint_S (v \partial \bar{u}_i / \partial x_j) N_j dSwb \right)$:

Drag force terms for different plants and roughness arrangements with different geometrical and flow conditions and their parameterization in CFD models are still being actively researched by several research groups (Nikora, 2009). In a simple modelling case, the drag force terms Fd_i can be parameterized by the conventional drag force equation (e.g. Naot et al., 1996; Nepf, 1999; Nikora et al., 2007b):

$$Fd \approx -\frac{1}{2}C_d S_f A_p |\langle \bar{u}_i \rangle| \langle \bar{u}_i \rangle \tag{5.14}$$

where C_d is the drag coefficient corresponding to a single plant or roughness element, S_f is a sheltering factor arising from the proximity of other elements, A_p is the averaged frontal projected area of the elements and $|\langle \bar{u}_i \rangle|$ is the resultant time-space averaged velocity. The composite term $C_d S_f$ is normally defined as the bulk drag coefficient.

Turbulent Reynolds stresses $-\langle \overline{u_i' u_j'} \rangle$

The turbulent Reynolds stresses in the momentum equation (5.12) can be calculated with the spatially-averaged $\langle k - \varepsilon \rangle$ turbulence model (Pedras & de Lemos, 2001; de Lemos, 2006):

$$-\phi \langle \overline{u_i' u_j'} \rangle = v_{t\phi} \left(\frac{\partial \phi \langle \bar{u}_i \rangle}{\partial x_j} + \frac{\partial \phi \langle \bar{u}_j \rangle}{\partial x_i} \right) - \frac{2}{3} \phi \langle k \rangle \delta_{ij} \tag{5.15}$$

where ϕ is the porosity, $\langle k \rangle$ is the spatially-averaged turbulent kinetic energy given by $\langle k \rangle = \langle \overline{u_i' u_i'} \rangle / 2 = \langle u_i' \rangle \langle u_i' \rangle / 2 + \langle \overline{u_i'' u_i''} \rangle / 2$ (Pedras & de Lemos, 2001), $\langle u_i' \rangle$ is the space-averaged temporal fluctuation, u_i'' is the spatial deviation in the time fluctuation, δ_{ij} is the Kronecker delta function and $v_{t\phi}$ is the turbulent eddy viscosity for the porous medium. In equation (5.15), the turbulent Reynolds stresses can be partitioned as $-\langle \overline{u_i' u_j'} \rangle = -\left(\langle u_i' \rangle \langle u_j' \rangle + \langle \overline{u_i'' u_j''} \rangle \right)$ where the first-term represents the time fluctuations and the second term is associated with the turbulent dispersion in the porous medium due to both time and spatial fluctuations (Pedras & de Lemos, 2001; Nikora et al., 2007b). The components $-\langle u_i' \rangle \langle u_j' \rangle$ and $-\langle \overline{u_i'' u_j''} \rangle$ represent the large-scale shear turbulence and small-scale wake turbulence due to these different physical mechanisms within the porous medium (Nikora et al., 2007b).

The turbulent eddy viscosity for a porous medium can be expressed in the same way as the Kolmogorov-Prandtl expression (Pedras & de Lemos, 2001):

$$v_{t\phi} = c_\mu \frac{\langle k \rangle^2}{\langle \varepsilon \rangle} \tag{5.16}$$

where $\langle \varepsilon \rangle$ is the spatially-averaged turbulent dissipation rate and c_μ is a constant. The spatially-averaged turbulent kinetic energy $\langle k \rangle$ and spatially-averaged turbulent dissipation rate $\langle \varepsilon \rangle$ quantities are determined from the spatially-averaged $\langle k - \varepsilon \rangle$ turbulence model equations (Pedras & de Lemos, 2001; de Lemos, 2006) with an parameterisation suggested by Shimizu & Tsujimoto (1993) and Naot et al. (1996) in the context of the conventional drag force equation (5.14) is written as:

$$\phi\langle\bar{u}_i\rangle\frac{\partial\langle k\rangle}{\partial x_i} = \frac{\partial}{\partial x_i}\left[\left(v+\frac{v_{t\phi}}{\sigma_k}\right)\frac{\partial\phi\langle k\rangle}{\partial x_i}\right] + P_k + G_k - \phi\langle\varepsilon\rangle \qquad (5.17)$$

$$\phi\langle\bar{u}_i\rangle\frac{\partial\langle\varepsilon\rangle}{\partial x_i} = \frac{\partial}{\partial x_i}\left[\left(v+\frac{v_{t\phi}}{\sigma_\varepsilon}\right)\frac{\partial\phi\langle\varepsilon\rangle}{\partial x_i}\right] + \frac{\langle\varepsilon\rangle}{\langle k\rangle}\left[c_{1\varepsilon}(P_k + G_\varepsilon) - c_{2\varepsilon}\phi\langle\varepsilon\rangle\right] \qquad (5.18)$$

where $c_{1\varepsilon}$, $c_{2\varepsilon}$, σ_k and σ_ε are empirical constants, P_k is the production of turbulent kinetic energy and G_k and G_ε represent the extra contribution arising from the material within the integration volume in equations (5.17) and (5.18) respectively (i.e. convective transport and production associated with spatial deviations of flow quantities). These extra terms are given as a function of the drag force F_i (Shimizu & Tsujimoto, 1993; Naot *et al.*, 1996) as:

$$G_k = \eta_k F_i\langle\bar{u}_i\rangle; \quad G_\varepsilon = \eta_\varepsilon F_i\langle\bar{u}_i\rangle \qquad (5.19)$$

where η_k is the efficiency of the production of turbulent energy and η_ε is the efficiency of the dissipation of turbulent energy. For the non-porous region above the roughness layer where $Fd_i = 0$ and $\phi = 1$, equations (5.17) and (5.18) resemble the standard $k - \varepsilon$ turbulence model. In present practice, the closure coefficients for the spatially-averaged $\langle k - \varepsilon \rangle$ turbulence model are assumed to be the same as the standard $k - \varepsilon$ turbulence model ($c_\mu = 0.09$, $c_{1\varepsilon} = 1.44$, $c_{2\varepsilon} = 1.92$, $\sigma_k = 1.0$ and $\sigma_\varepsilon = 1.3$). However, the applicability of these standard coefficients needs further investigation. For the efficiency coefficients η_k and η_ε, independent values can be selected as suggested by Shimizu & Tsujimoto (1993). However, Naot *et al.* (1996) and López & García (2001) both argue that the value of η_ε has to be dependent upon the value of η_k in order to maintain equilibrium between the production and dissipation of turbulent kinetic energy and remove a mathematical inconsistency. It should also be noted that the values of η_k and η_ε given by López & García (2001) are the ones which are consistent with the theory of flow through porous media (de Lemos, 2006).

5.7.4 Measurements for model setup and validation

In order to validate the equations being solved within a CFD model and improve model parameterisation, there is a need to acquire data from carefully controlled flume experiments, rather than field cases, which cover a wider range of flow and vegetation conditions. For setting up a CFD model, detailed measurements of flow domain (bed topography and water surface), bed roughness characteristics, plant characteristics (morphology, frontal projected area and porosity) and discharge are needed for wide range of flow conditions. For validation, particularly in flume experiments, it is essential to make simultaneous, spatially distributed measurements of flow velocity and turbulent kinetic energy both within and around plants as well as measuring the water surface variation. Measurement systems – Acoustic Doppler Velocimetry (ADV), Laser Doppler Anemometry (LDA) and Particle Image Velocimetry (PIV) can be used to measure flow field around plants and within plants, Hot-Wire Anemometry

(HWA) can be used (See also section 2). Parameterisation of drag coefficient can be either measured from experiments or derived from model calibration (e.g. Naden *et al.*, 2004).

5.7.5 Using field data for validation and parameterisation of numerical models

The choice of field locality, laboratory facility and model type need to be determined by the problems and issues to be addressed. In a natural system, biological organisms are adapted to specific niches and habitats, according to the species' physiological requirements. A physical model can only generate realistic data if it succeeds to mimic at least the key factors in the field system. Natural systems are experiments with a beginning, middle and end. Animals and plants colonize environmental niches successively and gradually until a climax community is established. It is therefore important to know whether the studied ecosystem is a climax community or a transient community. Field measurements are affected by a number of unknown factors which cannot be controlled; these can be minimised in the laboratory. It is also crucial to choose those parameters that are measurable. Some processes in the natural system are slow and therefore difficult to monitor, others may not be easy to observe for other reasons, such as depth, distance or the lack of appropriate monitoring equipment. In these cases, indirect measurements may be the only available mode of observation.

The specific methods used to validate a numerical model based on measuremed data, either from the field or the laboratory, depend on the model requirements and the quality and extent of data that can be collected. A key concept is the **observability** of the model (Kailath, 1980), which depends both on the model and the set of data available. If the model is observable, it means that the data set contains enough information for the entire model state to be estimated. For linear systems, observability can be determined using a simple mathematical criterion. For certain classes of nonlinear systems, for which no simple criterion exists, one might still be able to fulfil conditions that guarantee observability (Hwang & Seinfeld, 1972; Gauthier & Kupka, 1994). Still, for systems with distributed states, such as a hydrodynamic model, the typical situation will be that observability cannot be fully achieved. Methods based on optimal control theory can be used to find model states approximating the observations in such cases (Le Dimet & Talagrand, 1986).

Relevance and measurability may guide the researcher in his/her choice of environment when it comes to field studies, e.g. having the right biological species available and operating within a realistic experimental time frame. In the case of a physical model system, it has to be decided if the primary interest is to study fluid processes in relation to one particular animal or plant species or its internal interactions in an entire natural system.

The Kalman filter (Jazwinsky, 1970) can be applied for model correction and **parameter estimation using field or laboratory data** in linear systems. For nonlinear systems there are several generalizations of varying numerical complexity, such as the Extended Kalman filter, Unscented Kalman filter (Wan *et al.*, 2002) and Ensemble

Kalman filter (Evensen, 1994). In all variations, the state and parameter estimation can be done recursively – time step by time step – and can therefore be used in real time. The algorithms weigh measurement uncertainty against model reliability, thereby making it possible to suppress measurement noise.

5.8 SUMMARY CONCLUSIONS AND FUTURE CHALLENGES

This chapter highlight the need for greater interaction and collaboration between hydraulic engineers and ecologists. Combining the expertise of researchers from different disciplines is essential to improve the realism of physical models both in terms of their simulation of simple and complex flow and wave environments and the incorporation of plants and animals into those models. Through improved physical modelling that captures both the hydraulic and biological processes, we can develop a better understanding of the interactions between biota and their environment and the effect of the environment on the biota.

This chapter highlight the following key areas in which physical models of ecological experiments should develop: (i) the effect of vegetation and animals on modifying flow fields and (ii) the effect of plants and animals on patterns and rates of sediment erosion and deposition. Both these areas of research have important consequences for hydraulic engineering and ecology. Combining ecology and hydraulics introduces conflicting scaling requirements for reproducing representative flow and wave fields versus the limitations for scaling ecological components of the system which needs to be carefully considered at the experimental design stage. Inevitably simplifications are required for successful physical modelling, therefore it is important that physical models are considered in conjunction with field measurements and numerical modelling to broaden the applicability of the results obtained. There is no single set of criteria that can be defined for incorporating either plants or animals in physical models, therefore experiments need to be designed carefully around the husbandry requirements of the organism being studied and the experimental aims.

Chapter 6

Composite modelling

Herman Gerritsen & James Sutherland

João Alfredo Santos, Henk van den Boogaard, Sofia Caires, Rolf Deigaard, Martin Dixen, Conceição Juana Fortes, Jørgen Fredsøe, Paula Freire, Marcel van Gent, Xavier Gironella, Joachim Grüne, Nicolas Grunnet, Palle Martin Jensen, Rute Lemos, Maria da Graça Neves, Charlotte Obhrai, Filipa Oliveira, Tiago Oliveira, Hocine Oumeraci, Artur Palha, Liliana Pinheiro, Reinold Schmidt-Koppenhagen, Ulrike Schmidtke, Maria Teresa Reis, Manuel Marcos Rita, Francisco Sancho, Joan-Pau Sierra, Isaac Sousa & B. Multlu Sumer

6.1 INTRODUCTION

Hydraulic studies have been traditionally undertaken with physical models, which reproduce flow phenomena at reduced scale with dynamic similarity. Today, numerical models are increasingly being used in place of physical models. These models rely on mathematical descriptions of complex turbulent processes and boundary conditions but can be cheap and versatile. Physical and numerical models both have their strengths and weaknesses (van Os *et al.*, 2004) and their merits must be compared to the benefits of theoretical analysis (desk studies) and measurements made in the field.

A single tool cannot adequately reproduce the complex processes involved in coastal problems and thus replace all the others. Combining tools can add value, but also cost. In most cases the boundary conditions for a physical model come from a regional, numerical model. The models are run in sequence from large scale to small scale and frequently all the runs of one model are completed before the next (smaller area) model is run. Compromises are made with the validity of the models and about the (limited) amount of information that is transferred from one model to another. In the infrequent occasions when physical and numerical models of the same structure and beach are run, differences in the results are normally seen. The differences may be due to a number of factors, such as the poor description or parameterisation of processes in the numerical model, the use of simplified boundary conditions or idealised structures (possibly as designed, rather than as exists) in physical and numerical models. It is not always possible to decide which error is most important and to decide which model to believe.

This chapter describes techniques for combining physical and numerical modelling in order to improve the physical modelling infrastructure. Composite modelling can lead to different forms of improvements: being able to model problems that cannot be modelled by either physical or numerical modelling alone; increasing quality at the same cost or obtaining the same quality at reduced cost, reducing uncertainty at the same cost and realising that uncertainty reduction is also a quality issue. Composite modelling in the field of coastal modelling is a rather new approach, and relatively little has been published in the literature. This chapter will outline and summarise the eight composite modelling study cases that were conducted in the HYDRALAB CoMIBBS project.

The aim of CoMIBBS was to improve the service provided by hydraulic laboratories by developing techniques and good practice guidelines for composite modelling, which is the integrated and balanced use of physical and numerical models. For this, composite modelling techniques were developed and tested for four generic classes of seabed/fluid/structure interactions:

In the CoMIBBS project the four generic classes of seabed/structure interactions were:

- Shore-parallel structures
- Shore-permeable structures
- Jetties and groynes
- Piles and spheres

6.2   THE CONCEPT OF COMPOSITE MODELLING

The definition of Composite Modelling used in HYDRALAB is: *"Composite Modelling is the integrated and balanced use of physical and numerical models"*. This definition is rather wide and is based on an inclusive approach. Within the consortium of hydraulics research laboratories, university departments and other organisations co-operating in the HYDRALAB consortium, the links between the four building blocks of our knowledge on hydraulic processes and behaviour are often visualised in the following diagram (Figure 6.1).

The numerical and physical models encapsulate (part of) our hydraulic knowledge, theory and application experience and so make this objective, transferable and verifiable. The same is true for theoretical analysis and the description and interpretation of field experiments. Each building block links with the other three with a two-way flow of information and influence. As an example, field experiments influence theoretical analysis, while the results of theoretical analysis often suggest further field

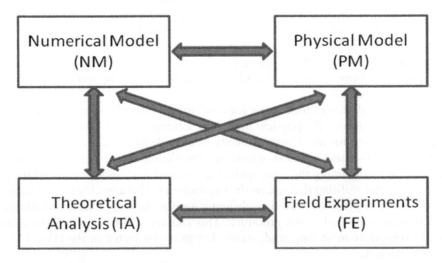

Figure 6.1 Links between the four elements of our research methodology or building blocks of our knowledge.

experiments. Physical model data are used to develop and calibrate numerical models, while theoretical analysis and numerical model results often lead to further physical experimentation to better understand complex process interactions and so on.

In this diagram, the building blocks or modelling tools are applied serially. In composite modelling, however, there often is such a strong interaction between field experiments and theoretical analysis, or between theoretical analysis and numerical modelling that one may speak of the integrated application of these building blocks.

In the HYDRALAB community the following ideas and expectations are associated with composite modelling:

- Combines the best out of both physical and numerical models for a given problem. Each can be applied to a geographical area or at those scales where it performs best, and so lead to an overall better simulation of the relevant processes.
- Will lead to improvements to the modelling infrastructure.
- Provides more quality (higher accuracy, reduced uncertainty) results at the same cost.
- Increases cost effectiveness for a given problem characterisation.
- Higher capabilities to model more complex problems which individual physical models or numerical models cannot.
- Upgrading hydraulic modelling to a new generation.

6.3 COMPOSITE MODELLING TECHNIQUE CASE STUDIES USED IN CoMIBBS

6.3.1 Nesting of a detailed physical model within a regional numerical model (DHI)

In this project, the conditions for bypass of sediment and associated sedimentation were investigated by application of composite modelling. A generic example was considered: a harbour on a straight coastline exposed to oblique waves. The protective works of the harbour consist of two curved rubble mound breakwaters located inside the surf zone.

A coastal harbour blocks the littoral drift causing accretion and erosion at the up – and downdrift side, respectively. Eventually a significant part of the littoral drift may bypass the harbour. In this case it is important to minimise the deposition of the passing sediment in the harbour basin and the harbour mouth and to maximise the water depth at the harbour entrance. Different measures have been proposed to counter sedimentation problems associated with sediment bypass. A solution is to design the protective works of the harbour to promote the bypass of sediment by making the harbour mouth face the incoming waves and streamlining the breakwaters. This increases the flow velocity and transport capacity past the harbour mouth due to flow contraction.

Composite modelling was conducted by combining a physical model in a shallow water wave basin with detailed area modelling of the fields of the waves, the current, the sediment transport and the resulting morphological evolution of the bed. The resulting coastline evolution was further simulated by a regional coastline model (a one-line model) based on a littoral drift model.

A physical scale model will by nature be very local, simply because in order to minimise the scale effects, the scale of the model should be as large as possible. Composite modelling was therefore carried out to combine the local physical model and regional numerical models in order for the two types of models to complement each other. Two different approaches for combining the local physical and regional numerical models have been considered.

Summary of composite modelling techniques and results

Transformation of forcing conditions to model domain

This work did not include analysis of field conditions to determine representative forcing conditions or design conditions. For a study for an actual site this is an integral part of the study, and for a physical model investigation the combined use of numerical and physical models can lead to an optimised scale of the physical model and improved representation of the forcing conditions in the physical model. Rarely the forcing conditions are known in the local area defined by the physical model, but can be transformed to the physical model domain by use of numerical models. The procedure is briefly outlined in the following:

Waves

Typically wave conditions are determined from field measurements (a wave buoy or a bottom mounted instrument) or from a regional or global hindcast model. In order to perform a realistic physical model, the wave conditions in the local domain and specifically at the boundaries of the physical model where the forcing is to be applied, must be determined. A regional numerical wave model can be used for transformation of the wave conditions to the local physical model area. To be suitable for the transformation the wave model includes the processes of depth refraction, energy dissipation due to breaking and the bed boundary layer and (depending on the extent of the regional model area) energy input from the wind.

Different methods are available for establishing the wave climate in the local area. The wave climate can be analysed in the regional data (the data point in the hindcast model or the buoy location) in the form of data for wave rose(s) or extreme events with recurrence periods of 10, 50 or 100 years. Each condition in the regional wave rose or the extreme events are then transformed to the local area and the local wave statistics are determined. Alternatively, a full time series in the regional data point is transformed, and the local climate is then determined by analysis of the time series in the local area. The latter method will often be preferable, especially for determining design conditions. Figure 6.2 illustrates an example: the wave rose represents the wave climate in the red point at the boundary of the regional area and have to be transformed to the boundary of the domain of a hypothetical physical model indicated by the red rectangle.

Currents

Similarly, a numerical flow model can be used to generate boundary conditions for a physical model from regional data. This can be done by calculating the discharge

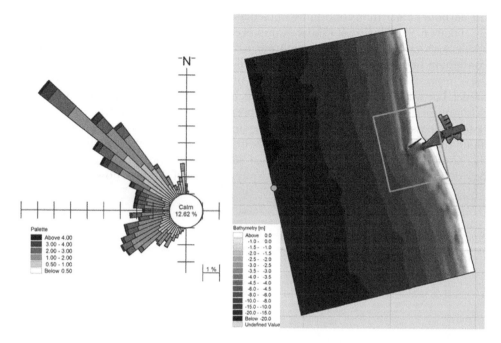

Figure 6.2 An illustration of a wave rose (for the red point) to be transformed to the boundary of a hypothetical physical model covering the area indicated by the red rectangle.

through a tidal inlet at a boundary of a physical model from the varying tidal elevation or by transforming measured current velocities at a location away from the physical model domain to the boundary of the physical model.

Derivation of internal boundary conditions for regional coastline model from detailed physical experiments

In DHI's study of a coastal harbour, the deposition of sediment in the harbour basin was determined in the physical model. Simulations were made with a coastline model to determine the regional coastline evolution caused by the blocking of the sediment by the harbour and the sedimentation in the harbour basin. The blocking of the longshore transport and the rate of sediment deposition in the harbour are internal boundary conditions for the regional coastline model and were specified on the basis of the results and observations from the physical model experiments. In this case the conditions for the regional model were thus obtained on the basis of results from the local physical model. Simulations were made for different amounts of sedimentation in the harbour and with different assumptions regarding how the sediment that bypasses the harbour is distributed along the coastline downdrift of the harbour.

In the physical model, the sedimentation in the harbour basin corresponded to approximately 5% of the sediment passing the harbour. The sediment that passed the harbour remained at an offshore position for a considerable distance downstream of the harbour rather than continuing immediately to travel in the surf zone.

Figure 6.3 and Figure 6.4 show examples of coastline simulations. Figure 6.3 shows results with varying rates of sedimentation in the harbour but with the passing sediment continuing immediately as littoral transport downstream of the harbour. The green curve represents the sedimentation of 5% of the sediment observed in the physical model, but was not realistic as it does not include the effect of the sediment remaining offshore. In Figure 6.4, simulations have been made for sedimentation of 5% of the sediment passing and with the bypassed sediment distributed over different distances along the coastline downdrift of the harbour. The erosion of the coastline downdrift increases significantly when the bypassed sediment remains offshore at a distance before reaching the coast. The size of the physical model did not allow determination

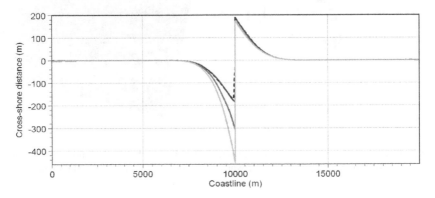

Figure 6.3 Simulated coastline evolution for different rates of sedimentation in the harbour, longshore sediment transport from left to right, light grey: 100% of the sediment passing the harbour is deposited in the harbour basin, medium grey: 50%, dark grey: 5% and black broken line: 0% sedimentation.

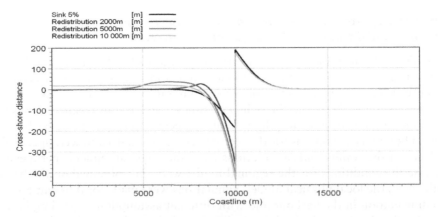

Figure 6.4 Simulations of the coastline evolution with 5% sedimentation in the harbour basin, black: 95% of the sediment passes directly to the downdrift coast, dark grey: the bypassed sediment is distributed over a distance of 2000 m downdrift the harbour, mid grey: over 5000 m and light grey: over a distance of 10,000 m downdrift the harbour.

of the actual distance over which the sediment returned to the coastline, but from the numerical coastline model it was found that the maximum erosion was not very sensitive to this distance.

Conclusions from the composite modelling case study

The case study showed that composite modelling can be used to combine regional and local models. The numerical models can transfer the forcing conditions from locations far from the physical model domain to the boundary of the model or to locations within the domain. Detailed information obtained from the physical model experiments may in turn be transferred to a regional model as in the present study, where the modelling of coastline evolution was improved on the basis of findings from the physical model.

When conducting physical scale tests, the emphasis should be put on optimising the scale and model technique of the physical experiments. The physical model can be considered as a local model. Rather than attempting to increase the area covered by the physical model, composite modelling should be used to transfer regional data on the forcing to the boundaries of the physical model. In specific cases, it could be useful to upscale the detailed physical model results by using them as input to regional numerical models. The present case considered details of bypass of longshore transport at a coastal harbour which was introduced into a model for coastline evolution.

6.3.2   Reduction of uncertainties in physical modelling using a numerical model error correction technique (DH)

Sandy nearshore bottom profiles have a dominating effect on the waves reaching a coastal structure. These profiles can change significantly under severe storms and it is therefore desirable to take bottom uncertainty into account in physical scale modelling of the (most critical) wave loads. The bottom uncertainty significantly increases the complexity and cost, so is never taken into account in physical experiments. With numerical simulation, this can be done much more flexibly and efficiently, but unfortunately numerical model results are not as accurate as physical scale experiments. The use of composite modelling can lead to efficient and accurate determination of the most critical bedform for maximum wave loads, that is, a reduction of the bed uncertainty in physical modelling in which just one or two 'well chosen' bed forms are considered.

To investigate the benefits of composite modelling for such problems, we considered the 1D case of waves over a typical schematised foreshore with a bar, and a Low Tide Terrace (LTT) with unknown length and height (α_1, α_2). This reference bottom profile is an idealisation of the bottom profile in front of the Petten Sea defence (The Netherlands) and is characteristic for many situations. Van Gent & Giarrusso (2005) have shown that the shape (depth and length) of the LTT are the main determining factors. Therefore, several LTT configurations have been taken into account in this study together with specific variations of the hydraulic boundary conditions. The physical experiments were carried out in the Deltares Scheldt flume and the numerical experiments were made with the Boussinesq-type wave model TRITON.

The aim of the present CM case study was to determine or optimise the bed form parameters (α_1, α_2) that give rise to the most critical wave loads $(H_s, T_{m-1,0})$ at the toe of the structure or dike of a foreshore with a bar and a Low Tide Terrace (LTT) with unknown length and height (α_1, α_2).

The same problem was modelled using both Physical scale Model experiments (PM) and Numerical Modelling (NM). This allowed a one to one comparison of the results of both approaches in terms of wave height H_s, *spectral* wave period $T_{m-1,0}$ and water level h in all measurement locations, for a range of relevant boundary forcing $(H_s, T_{m-1,0}$ and $h)$. The composite modelling approach consisted of deriving a dedicated model for the systematic errors in the numerical model from a limited number of identical simulations with PM and NM. This error correction model was based on a neural network, which was trained on the results of the identical PM and NM simulations.

The hypothesis underlying the case study was that the uncertainty in critical bed forms in the PM can be reduced by determining the most critical bed form parameters (α_1, α_2) via efficient NM simulations in which (α_1, α_2) are systematically varied, and the corresponding wave loads are determined. Application of the error correction model to the NM results led to results of wave loads as function of (α_1, α_2). The corrected NM results will then have accuracy of near – physical model result quality. Therefore, the bed form parameters for which wave loads are maximum will also be accurately determined. A final PM experiment with the determined LTT (α_1, α_2) will determine / confirm the critical wave loads that were identified through CM.

Summary of composite modelling techniques and results

Two wave load formulations or targets L_T were considered. The first was the square root of the significant wave height $(\sqrt{H_{m0}})$ times the mean wave period, $(T_{m-1,0})$, leading to Target$_1$ defined by $L_T = \sqrt{H_{m0}} \cdot T_{m-1,0}$. This quantity represents a measure for the damage of a construction at the toe due to wave attack. The other wave load formulation Target$_2$ read $L_T = H_{m0} \cdot T_{m-1,0}$, which provides a measure for the amount of wave overtopping at the construction. The H_{m0} and $T_{m-1,0}$ in these two wave load formulations Target$_1$ and Target$_2$ referred to their values at the toe of the coastal structure.

In both the PM and NM, five bedforms were tested for 12 hydraulic BCs, which led to 60 data points for which both PM and NM results were available. The PM results were assumed to represent the truth. By plotting the results of PM and NM in the form of the relative error (NM-PM)/PM for the 5 bedform cases simulated against various parameters such as ak^2, α_1, α_2, the error showed a systematic dependence on ak^2. This was used to guide the development of a NN to model the error, i.e. to fit a best curve through the data points for the relative error.

Three NN architectures of MLP (Multi Layer Perceptron) type were tested and calibrated in terms of best fitting the experimental results (60 data points: 12 hydraulic conditions for 5 test bed forms). Figure 6.5 shows the calibration results for the three MLP versions of the error models. The 2.5 parameter MLP gave the best results, while still being robust.

The application of this NN based error correction removed or at least strongly reduced the NM error and created PM-like quality results (Figure 6.6 a, b). For the next step, an (efficient) sensitivity analysis with the NM was performed in which the

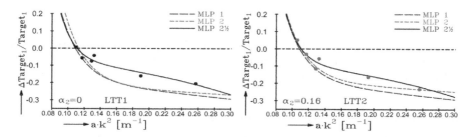

Figure 6.5 Results of the calibration of the three MLP versions of the error models: the 1-parameter MLP version (ak², long dashes), the 2-parameter MLP (ak², α_2, short dashes) and a 2.5 parameter MLP (ak², α_2, and water depth h, solid line). The left and right frames show the results for two different LTT experiments.

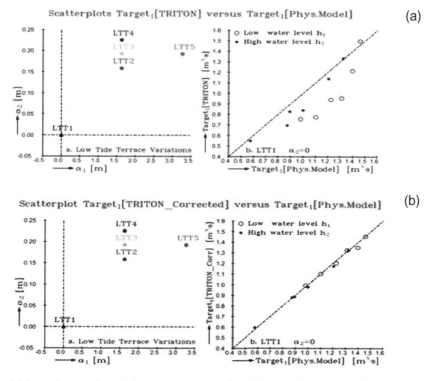

Figure 6.6 (a, b) Scatter plots of the relative numerical model error (a) before application of the error correction (upper frame), and (b) after application of the correction (lower frame), for the case of LTT1 and Target₁ (other cases showed similar results).

wave-loads at the toe (Targets) were determined as a function of (α_1, α_2) by systematic variation of the shape parameters (α_1, α_2) of the bed form. The NN-based error correction was applied inversely to the NM-wave-load function to derive a PM-like quality wave load dependency on (α_1, α_2). The maximum of this curve gave the (α_1, α_2) pair for which wave loads are most critical. Since the calibration experiments showed very little dependence on parameter α_1, the sensitivity analysis was restricted to varying α_2 (see Figure 6.7).

A final PM experiment would have been useful for confirming these critical wave loads. However, in this case the maximum was almost reached in one of the five cases tested so this verification was not needed.

Conclusions from the composite modelling case study

Analysis of the differences in wave loads predicted with the physical model and those predicted with the numerical model, showed a systematic dependency on the off-shore hydraulic conditions, and the shape/size of the low tide terrace. This systematic behaviour allowed the construction of a neural network-based error correction procedure for the numerical model, assuming the physical scale experiments as benchmark for accuracy. With this error correction, the accuracy of numerical model predictions were upgraded to a quality that closely approximated that of (corresponding, but not carried out) flume experiments.

As a result, identification of, and conclusions on, most critical situations (and/or other issues in design studies), were made much more accurately and reliably than on the numerical model results. Moreover, this combination of the numerical and error correction model (representing the present composite modelling approach) could significantly reduce the number of flume experiments to derive such critical conditions by PM only.

The essence of the case study is combining a flexible but not sufficiently accurate NM with an accurate but expensive PM to improve the PM approach. A key element in the CM is the error correction procedure. In principle, this concept can be applied in a very general way. The key to the success of the present composite modelling was the prior analysis of the model deficiencies and conceptual physical/mathematical characteristics. The determined dependency of the numerical model error on ak^2 strongly guided the error modelling. It is not clear whether such guidance can always be found.

In the most general case investigated, the error depended on up to five parameters: the LTT shape parameters (α_1, α_2), water level h, and incoming wave parameters

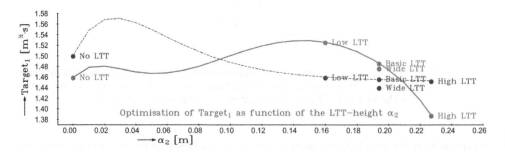

Figure 6.7 Results of the sensitivity analysis, c.q. optimisation of the wave load Target$_1$ $L_T = \sqrt{H_{m0}}$ $T_{m-1,0}$ as function of the height α_2 of the low tide terrace. Dashed/full line dependence on α_2 before and after application of NN based error correction.

$(H_s, T_{m-1,0})$; whilst in the present case we were able to effectively reduce the sensitivity analysis to only one for α_2, more computing power or parallelisation will be needed to do a sensitivity and optimisation analysis in a 3D or 5D space. Within the case study, the numerical model appeared less efficient than expected, since a finer spatial grid resolution was needed for the cases with largest wave steepness. Nevertheless, this application of composite modelling was successful in terms of reducing effects of uncertainty in bedform set up.

6.3.3 Determination of optimum physical model scale (LNEC)

Wave breaking and sediment transport near the vicinity of coastal structures are very complex phenomena. The accuracy of modelling these phenomena, through laboratory experimentation, depends very much on the model scale used. Therefore, it would be desirable to make physical model tests at different scales to evaluate their effects on these phenomena. However, this is a lengthy and expensive procedure and thus a combination of physical modelling and numerical modelling seems to be the best compromise for achieving an optimised simulation. The numerical models will be used to guide the choices for more efficient physical modelling.

The problem under study by LNEC (LNEC, 2008), was the influence of the physical model scale in (i) the simulation of wave propagation on coastal defences, in particular where the wave breaking phenomena plays an important role, and (ii) the beach profile evolution in front of alongshore structures such as seawalls/revetments.

In relation to hydrodynamics, wave breaking close to the structure is a key point in the simulation of wave interaction with coastal structures. Since the wave height at breaking and the breaker position vary with the physical model scale, to evaluate such scale model effects, one would have to perform a large number of experiments at different scales. Therefore, the methodology proposed is to use a Numerical Model (NM) to define the smallest scale to be used in Physical Model (PM) tests in a flume that complies with a pre-set level for model scale effects in the wave propagation on a plane slope that ends on a coastal defence. Conversely, scale model results were used to fine tune the numerical model parameters in order to better describe the flow (especially wave breaking, i.e., wave breaking position and height) in the model geometry.

In relation to sediment dynamics, finding the most suitable scaling laws for sediment dynamics in the vicinity of coastal structures needs more investigation. Scaling of the particle size is the key problem in physical simulation of sediment transport, primarily when the prototype sediment is sand. Therefore, this was carried out in experiments of the beach profile evolution under storm and mild wave conditions at two laboratory scales. These were combined with Numerical Modelling (NM), which mainly assisted the design of experiments, and evaluation of strengths and weaknesses with respect to scaling problems.

The overall objective of this study was to establish a methodology, based on the combined use of physical and numerical models, to determine the optimum physical model scale for the hydrodynamics (focusing on wave breaking height and position) and sediment dynamics of propagating waves reaching coastal defences (seawalls, revetments, dikes, dunes or submerged breakwaters).

Summary of composite modelling technique and results

In terms of the hydrodynamics the following steps were performed:

- Definition of Case Study – Establishment of the foreshore area of "São Pedro do Estoril" sea defence as the study area;
- Physical Model Experiments – Physical model tests were performed in different scales to calibrate and validate the numerical model and to enable the evaluation of physical model scale errors associated to wave breaking height and position;
- Numerical Model Calculations – A calibration of the numerical model COUL-WAVE (Lynett & Liu, 2002) was performed with the physical model results obtained. The numerical model errors (obtained by comparison with the physical model results) were determined;
- Validation of the Composite Model
  - o   Assessment of the Error Evolution – For wave conditions not used in the calibration, numerical calculations with COULWAVE model were performed in a systematic way for the case study and for the model scale range that was considered to be feasible in LNEC's wave flumes. This enabled an evaluation of the numerical errors due to scale effects in wave breaking height and position;
  - o   Using a data subset of "Numerical Model Calculations", a first assessment of the goodness of the estimates of the numerical model errors due to scale effects was made for the scales tested in the physical model.

In this study, it was determined that the numerical and physical model results for scale 1:10 do not suffer from scale effects. Therefore, after choosing the study area, a physical model was constructed representing a 1:20 beach slope ending on a 1:1.5 seawall. Two bottom slope and structure configurations were considered: Case A, where the depth at the toe of the structure was 10 m or 11.5 m; and Case B, where the depth at the toe of the structure was 20 m or 23 m.

The study focused on the complex physical processes involved in the breaking zone. Tests were undertaken using 1:10, 1:20, 1:30, 1:40 and 1:60 physical model scales, (each case was represented with three different scales), involving different incident wave conditions (regular and irregular waves). The free surface elevation measurements along the flume obtained for the three physical model scales for $t = 12$ s, $H = 4$ m and $h = 11.5$ m (Case A), were used to calibrate COULWAVE and to evaluate the errors of the numerical and physical models due to scale effects. So, the numerical results, in terms of free surface elevation for those conditions, were obtained along the domain and in particular at the physical model wave gauge positions. Based on those values, significant wave heights (experimental and numerical) were calculated as well as the wave breaking height, Hb, and wave breaking position, Lb. The numerical model parameters were tuned so that the numerical model errors associated to those quantities were small.

The validation of the proposed composite modelling methodology consisted of comparing the scale effect error for wave breaking height and position obtained with the numerical model for two selected test cases not used in the calibration of the numerical model (Case B – $t = 12$ s, $H = 2.0, 4.0$ m, $h = 23$ m) to the corresponding values obtained from physical model measurements. The work performed was:

- Use of the numerical model COULWAVE for Case B tests, considering a range of model scales from 1:10 to 1:60;
- Estimation of the numerical errors due to scale effects considering the scale 1:10 as the reference scale;
- Comparison between the physical and numerical errors due to scale effects, Table 6.3.1. These errors are defined as:
  - o Error in the physical model (Error PM) = (physical value at given scale – numerical value at scale 1:10) / numerical value at scale 1:10
  - o Error in the numerical model (Error NM) = (Numerical value at given scale – numerical value at scale 1:10) / numerical value at scale 1:10.

The errors due to scale effects obtained by the physical and numerical models are consistent for Lb, but not for Hb. This leads to the conclusion that the numerical model employed in this test is not the most adequate for this sort of studies since it gives almost no scale effects for model scales where such effects are expected to be important. Such a conclusion does not invalidate the proposed methodology since the same physical model data can be used with a better numerical model.

Sediment dynamics

The composite modelling approach presented in Figure 6.8 was followed in order to find the best small-scale physical model to simulate beach profile evolution, in a realistic way. Models at two different scales were tested: a large scale (1:6) and a small-scale (1:13).

At the initial stage, the numerical model Litprof (DHI, 2007), a deterministic 2D-vertical cross-shore morphological model, was applied to assist the design of the large-scale experimental set-up, including: i) the estimation of the minimum thickness of the sand layer that should be used in the experimental set-up in each flume; and ii) the analysis of the effect of the geometry of the offshore part of the profile on the morphological evolution of the total profile. The experimental set-up was based on a field site, Buarcos beach, located on the Atlantic coast of Portugal, and the initial geometry was a planar profile with 1:20 slope, in front of a reflective structure.

Table 6.3.1 Case B, $h = 23$ m, $t = 12$ s, $H = 2$ m. Physical and numerical errors due to scale effects.

	Lb		Hb	
Scale	Error_PM	Error_NM	Error_PM	Error_NM
1:20	2.9%	2.9%	6.1%	0.1%
1:40	2.9%	2.9%	18.2%	1.2%
1:60	2.9%	0.1%	27.1%	3.9%

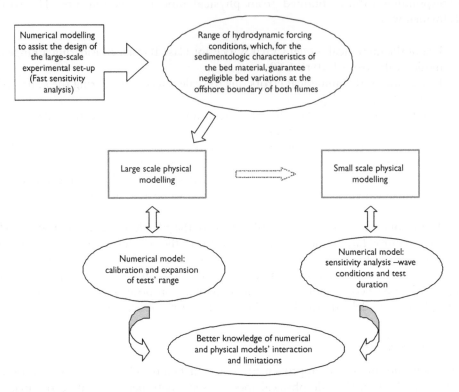

Figure 6.8 Schematic view of the composite modelling approach.

Physical tests were performed for erosive and accretive conditions in a large and a small scale wave flume. The wave conditions for small-scale tests were based on those simulated at the large-scale flume, considering the Froude scaling-law, with the scale factor $n = 2.22$. Wave periods and test durations were scaled according to $\sqrt{n}$. Following Van Rijn (2007a), the suspended-load scaling law, $n_{D50} = n_h^{0.56}$, was considered for the sediments scaling, where n_{D50} and n_h are, respectively, the scaling relation of sediment median diameter (D_{50}) and of the depth. The sediment used was silica sand (Sibelco-SP30) with $D_{50} = 400$ μm and geometrical deviation, defined as $(D_{80}/D_{16})^{0.5}$, of 1.35 in the large scale flume; $D_{50} = 250$ μm and geometrical deviation of 1.22 in the small scale flume. According to the above, the laboratory tests represented scaling conditions of a prototype beach with $D_{50} = 1.1$ mm, at scales of ~1:6 and ~1:13 for the large and the small-scale flumes, respectively. Moreover, all hydrodynamic conditions were such that mobility numbers assured the same bed-regime at all scales and in the prototype (Ribberink, 2007). For each test the data acquired included bed profiles, surface elevations, 3D velocities, sediment porosity after each test and beach profile photographs.

In the large scale physical-model tests, the initial planar profile developed to a typical barred-beach profile during erosive conditions. On the contrary, onshore sediment transport was observed during accretive waves, promoting the landward migration of the bar. When wave-structure interaction was simulated due to increased water level, significant erosion near the seawall was observed under erosive conditions. These results showed that the water level in front of the seawall controlling the wave-structure interaction, was determinant for profile development, particularly near the structure.

The numerical model was then applied to simulate the large scale tests for calibration and validation purposes. Results showed that the numerical model could represent the beach profile configuration in a consistent way in terms of erosive conditions. However, under accretive conditions, the numerical model was not capable of simulating sand accumulation at the beach face due to limitations of the model in simulating swash processes.

At the small scale, physical model tests were initially carried out for a (predicted) erosive wave condition. This test yielded a beach profile without the features of erosion type, namely a barred beach, and thus the combined modelling strategy involved using the numerical model to perform a sensitivity analysis to the test parameters: incident wave height, wave period and test duration. Based on the numerical results, modifications of the initially planned wave input conditions were tested in the physical model. The experimental results from those new tests showed that a modification of the wave period was not relevant for the profile evolution pattern, whereas the test duration had a major influence in the bottom evolution.

The sensitivity analysis performed with the numerical model, and the lack of agreement between the small-scale numerical and experimental results, confirmated that the morphological behaviour of the beach profile at the small-scale is not a scaled-down version of the corresponding behaviour at the large-scale.

Conclusions from the composite modelling case study

Hydrodynamics

In terms of the hydrodynamics, the numerical model COULWAVE helped in the design of the physical model tests and in the set-up of the measurement equipment. It was also able to predict the physical model errors due to scale effects associated with wave breaking position. However, the NM was not able to provide a correct estimation of the physical model errors due to scale effects associated with the wave breaking height. This was a consequence of numerical model limitations, especially in the breaking area and whenever the nonlinear and the reflection phenomena become significant in the wave propagation process. For certain occasions, the numerical model was applied for wave conditions outside the range of its applicability. Furthermore, the numerical model did not include surface tension and viscous effects that become important in wave breaking when H = 0.05 m or lower. The physical model results enabled the calibration of the numerical model, however, errors in the physical model measurements resulted in errors in the calibration of the numerical model parameter and, consequently, to errors on the composite model results.

The use of a more sophisticated numerical model, based for instance in the RANS equation, can provide a better simulation of wave breaking and so improve the composite methodology. Composite modelling was an adequate tool to study wave propagation as it significantly improved the outcome as compared to a single methodology.

It also improved the design and optimization of the physical model tests. In fact, once the best scale is defined (based upon the results of the evolution of the numerical error due to scale effects), the experiments can be planned and the measurement equipment can be better deployed in the flume, in order to get the most complete and accurate set of results. An increased range of conditions were validated for the numerical model from result contributions of the physical model tests. Composite modelling saves time and aids understanding of the hydrodynamics involved in propagating waves reaching coastal structures and the shortcomings of numerical models.

The limitations of the composite modelling are still inherent of the specific limitations (weaknesses) of each modelling component – physical and numerical. Limitations of the numerical model (COULWAVE) are related to its application outside the range of conditions for which the model was calibrated and to preview the wave height scale model errors. Another limitation is related with the error in the physical model measurements. These errors lead to further errors in the numerical model parameters, which were fine tuned based on them and, consequently, to errors in the composite model results.

The composite modelling technique is a flexible tool since it can use different numerical models for the physical phenomena under study, using the same methodology. The use of a more sophisticated model, based in RANS equations, like COBRAS-UC, (Lara *et al.*, 2006), would have provided a better simulation of the wave breaking, and in this way would have improved the composite modelling methodology.

Sediments

The composite modelling of the morphological behaviour of the beach profile evolution greatly improved the results and conclusions compared to using only one technique (either numerical or physical). Moreover, it optimized the use of physical modelling. According to the composite modelling strategy performed, the numerical model assisted the design of physical model tests, which in turn yielded results that allowed improvements in the use and confidence of the numerical model for a greater set of conditions. It was found that the water level in front of the seawall (that controls wave-structure interaction) determines beach profile evolution, particularly, the degree of beach erosion in front of the seawall.

Differences were observed in the success of the combined physical and numerical models to simulate the beach profile morphodynamics at the large and small scale. At the large scale, the numerical model was seen to be limited in reproducing beach recovery (under accretive conditions), but performed reasonably in simulating beach erosion (under storm waves). Both numerical and physical models results did not agree or represent the observed beach morphodynamical behaviour at the large-scale, probably due to an inadequate choice of scaling laws and sediment characteristics.

The initial composite modelling objectives were not completely achieved due to the failure in reproducing an erosive-beach behaviour at the laboratory small-scale tests. Therefore, an optimal (beach profile morphological) scale and scaling-law was not found. This enhances the difficulty in performing small-scale experiments, due to the variety of physical processes and scaling-laws related to sediment mobility and transport, and morphology evolution.

The combined modelling approach to study beach profile evolution performed herein can be generalised to other studies. However, despite a carefully planned methodological approach, we found the need to adapt it during the study. Therefore we can recommend that the use of combined modelling should be kept flexible, in order to make the best use of the strengths of PM and NM and allow adaptation to the particular study. It can also provide an answer when things go wrong.

Finally, although we tackled the complex problem of sediment dynamics scaling via a combined modelling methodology, we failed to properly understand the appropriate scaling relations, and in improving the physical modelling of beach profile evolution at small-scale flumes. However, the composite modelling improved the optimization of experimental set-up and the understanding of experimental results. It did not improve the weaknesses of the numerical model but allowed identification of their magnitude. We found that the limitations of our CM were mainly due to the specific limitations (weaknesses) of each modelling component (physical and numerical), which were not totally overcome.

6.3.4 Use of a numerical model to design a physical model (DHI)

The conditions for bypass of sediment and the associated sedimentation at a harbour on an open coast were investigated with a combination of physical and numerical modelling. The sediment transport was modelled by fine sand acting as a tracer over a hard concrete bed to illustrate the bypass of sediment and intrusion into the harbour. Further experiments were conducted with a moveable bed over part of the model domain. The physical model provided information on the flow, wave field, areas of sedimentation and mechanisms for the bypass of sediment. The numerical models were run in parallel to the physical modelling, simulating the detailed flow and wave patterns.

The composite modelling was based on the use of mathematical models to design and plan the physical model experiments. Numerical modelling was applied to provide boundary conditions for the physical model by transforming the forcing data from prototype scale to laboratory scale within the given physical restraints. In the current study, boundary data included waves and littoral current; however the use of numerical analysis could also have been extended to include water level variation such as storm surge. Upon numerically deriving adequate boundary conditions, limits for the model scale could be inferred from the limitations in wave height, taking into account laboratory restrictions (i.e. performance of the wave maker for a given water depth, spatial requirement for the generation of a fully developed wave-induced longshore current within the given basin size). The mathematical model was also used for the design of the physical model for the size of the harbour, the harbour location between the lateral sides of the basin and the seaward distance to the wave maker in order to ensure that the blocking effect was not causing problems in the physical model. In addition, numerical modelling allowed for an initial assessment of the conditions for sediment transport in order to ensure that sediment would in fact be mobilized in the down-scaled laboratory experiment.

The definition of the physical model layout was undertaken by considering the available facilities and optimized by the use of numerical modelling. The basin for of the physical model was one of DHI's shallow water basins at 35 m long, 25 m wide and 0.8 m deep. A 17 m long 3D wave maker was fixed along one side of the basin, generating irregular, multidirectional waves. Long-crested wave makers were also available. Each were 5.5 m long and generated irregular, unidirectional waves. The shallow water basin was also equipped with a pump system capable of supplying a maximum discharge of 0.2 m³/s.

An alongshore uniform bathymetry defined by constant slope of 1/25 was modelled. This slope characterises a dissipative beach and represents a typical slope on the offshore side of a breaking bar. In order to produce significant longshore sediment transport, waves were generated obliquely to the coastline at an angle of 25° at the offshore boundary of the model. The significant wave height (H_s) was set to 8 m, and the peak wave period (T_p) was 12 s. In the physical model, two of the wave makers were unidirectional, consequently they had to be rotated at 25° to the 3D wave generator system to create the desired offshore wave direction.

The balance between costs of physical scale models and numerical modelling has changed during the last 25 years with numerical modelling becoming comparably cheaper. The hypothesis is that when physical modelling is found to be the optimal solution, it can be advantageously combined with numerical modelling. The numerical model can be applied to optimise the layout and scale of the physical model and to determine the boundary conditions to apply in the physical model. This process of composite modelling should lead to the minimisation of laboratory effects and to a reduction of the time required for running in the physical model.

Summary of composite modelling techniques and results

Location of the harbour and inflow definition

In the hydraulic scale model, the harbour had to be placed far enough away from the inflow boundary, in order to obtain a good cross-shore distribution of the fully developed current. To determine this location, a numerical model including the physical constraints and boundary conditions described above was run against a periodic model representing the idealised uniform coast. The resulting wave fields are shown in Figure 6.9 and Figure 6.10 for the numerical modelling of uniform conditions and the laboratory conditions respectively. In both models, the current was extracted along cross shore profiles defined between 0 m and 2,500 m. Figure 6.11 indicates clearly that the current profiles from the uniform and the prototype numerical models fitted perfectly at 1,000 m. Thus the harbour entrance could be placed at this location. Moreover, the required inflow volume at the eastern boundary of the prototype model was found to be 10,000 m³/s, corresponding to the integrated flux at 1,000 m.

Scale between prototype and physical model and finalisation of the physical model

In the previous section, it was established that the wave driven current generated by waves characterised by a significant height of 8 m and a peak period of 12 s was fully developed after 2,500 m. Due to the dimensions of the physical basin, the physical model was consequently defined with a length scale of 1:100. According to Froude's

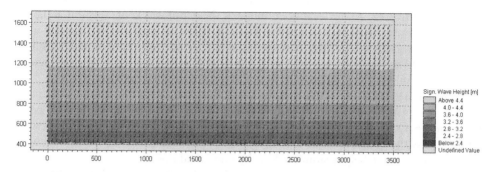

Figure 6.9 Wave height and direction field from a uniform (periodic) numerical model.

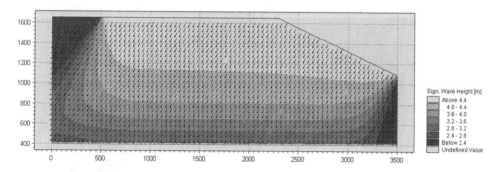

Figure 6.10 Wave height and direction field in the laboratory (without wave guide walls).

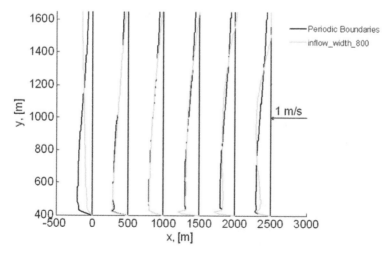

Figure 6.11 Current velocity distribution extracted along the x-direction from the uniform, periodic model (black lines) and the laboratory model (grey lines).

law, the corresponding time scale was 1:10. The final layout of the physical model is shown in Figure 6.12. At the offshore boundary, the depth was kept constant at 0.55 m to ensure no wave breaking occurred in front of the wave makers. Wave guide walls were placed on both sides of the offshore boundary. At the upstream model boundary, the inflow was generated by a pump with a discharge measured by an electromagnetic flow meter, while at the downstream boundary, water flowed out over a weir.

The geometry of the wave guides were determined in the numerical model in order to ensure wave conditions as close to the uniform conditions as possible. Figure 6.13 shows a comparison between the variation in the significant wave height across the coastal profile measured in the physical model by resistance wave gauges and the wave height simulated in the numerical one.

Wave-driven currents from the numerical model are presented in Figure 6.14. It was observed that the longshore flow had adjusted to near-uniform conditions

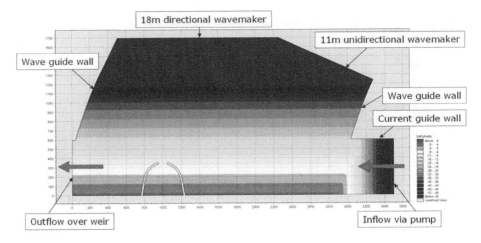

Figure 6.12 The physical model domain.

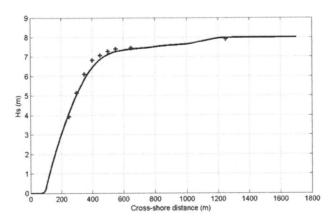

Figure 6.13 Comparison of the significant wave height predicted by the numerical model (solid line) and the experiments (crosses).

before reaching the harbour and the outflow had little effect on the conditions at the harbour. A comparison between the inflow magnitude in the laboratory and the numerical model at 3,000 m was established in order to make sure that the forcing at the boundary was similar for both models. The results are shown in Figure 6.15. The flow was measured by a Nourtec acoustic backscatter current meter at approximately 1/3 of the depth from the bed, while the numerical model represents the depth averaged flow velocity. Inflow from the experiments and the numerical model had the same patterns and both tended to increase into deeper water, whilst the numerical model tended to give too small velocities at the more shallow inner part of the profile.

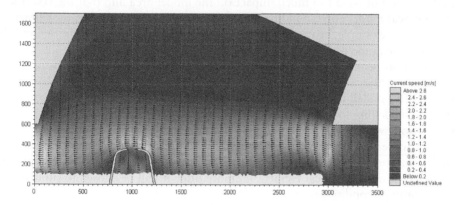

Figure 6.14 Wave-induced current magnitude and direction for the physical model domain simulated by MIKE21 HD, note that the wave guides following the wave rays at each end of the wave generator front are included in the model.

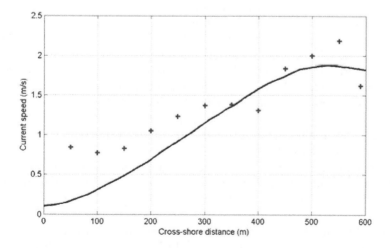

Figure 6.15 Comparison of the current speed at the inflow (x = 3,000 m) between the numerical results (solid line) and the physical measurements (crosses).

Conclusions from the composite modelling case study

In the case study, the numerical models were very useful in designing the set-up of the physical experiments. In particular, the numerical models were able to represent the conditions in the test basin well enough to make it possible to determine the maximum allowable model scale and optimal location of the harbour in the model basin. The physical and numerical models agreed so well that it would be possible to extend many of the physical model results to other conditions (e.g. wave direction, wave height and water level) by simulations with the numerical models.

Whenever a numerical model gives a realistic representation of the main features of a physical model test, it can be used to design and optimise the layout of the physical model. It can then also test different model configurations to ensure that the boundaries do not have too much impact on the model area and that the conditions in the physical model correspond to the prototype conditions. Finally, the numerical model can, for example, be used to design the in-flow and outflow conditions and the layout of wave guide walls.

Simulation results from the numerical models should be compared to measurements in the physical model. Numerical modelling is often found to be less costly than physical modelling. Therefore the physical model results may be extended by numerical modelling for a wider range of forcing parameters. This may also be done for conditions where scale effects in the physical model would be severe, for example under more gentle wave conditions where the model sand would be stable even though transport would occur in nature. The numerical model could then represent the conditions in the physical facility, but more often it will be chosen to simulate the actual prototype conditions.

6.3.5 Sensitivity analysis using skill scores (HRW)

Detached breakwater schemes interact with the near-shore hydrodynamics in a highly complex way. Schemes around the UK have commonly been built in a macro-tidal environment. Hence the waves approaching the breakwaters have evolved due to refraction, shoaling, non-linear interactions, reflection and dissipation associated with wave breaking. Accurate prediction of the seabed's response to the breakwater is important for the scheme designer and for coastal managers.

Detached breakwater schemes have been modelled using Physical Models (PM) and Numerical Models (NM), which both have their strengths and weaknesses (Sutherland & Obhrai, 2009) as summarised below:

- PM strengths: nonlinear processes; local scour and diffraction;
- PM weaknesses: scale effects;
- NM strengths: can model a set of breakwaters without scale effects;
- NM weaknesses: simplified processes, especially scour, diffraction and representation of the breakwater with no swash-zone processes.

We have attempted to combine the strengths and reduce the weaknesses by adopting a composite modelling approach, which has been defined as the integrated and balanced use of physical and numerical models. 'Integrated' implies the use of a physical and a numerical model, while 'balanced' implies an optimised flow of information between the models. 'Optimised' flow of information has been taken to mean sufficient but not excessive (giving a cost-effective solution).

Physical model

The physical modelling (Obhrai & Sutherland, 2009) was performed in a coastal wave basin with a single detached offshore breakwater in the centre of a bed of fine sand (with d_{50} = 0.11 mm) as shown in Figure 6.16. Five tests were run of the same case, only with changes to the representation of the water levels, currents and transmission of the breakwater. A laser scanner was used to measure the full physical model bathymetry, which was then transferred to the numerical model. An example model final bathymetry is shown in Figure 6.17 (left).

Numerical model

Numerical modelling was undertaken using the coastal area model PISCES (Garcia-Hermosa *et al.*, 2009) which used the finite element flow model TELEMAC, the third generation coastal area wave model SWAN and the sand transport module SAND-FLOW to simulate sediment transport, which drove changes in the bathymetry.

Figure 6.16 Physical model at HR Wallingford.

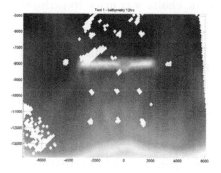
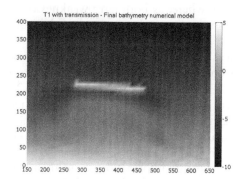

Figure 6.17 Physical model (left, dimensions in mm) and numerical model (right, dimensions in m) bathymetries at end of Test 1.

Numerical model simulations were performed of full-scale versions of the physical model, with variations in model set-up. An example final bathymetry is shown in Figure 6.17 (right).

The hypothesis tested was that a quantitative technique for optimising the information flow between models could be developed using skill scores (Sutherland et al., 2004a). The optimum information flow could be determined by calculating the skill of each model run. The skill scores could give a quantitative measure of the incremental benefits of adding additional processes or using more complicated boundary conditions (Sutherland & Obhrai, 2009). This would allow an optimal model setting to be chosen that gave a high skill score without using excessive resources.

Summary of composite modelling techniques and results

Skill scores

Skill is a measure of the accuracy of a prediction (compared to the correct outcome) relative to the accuracy of a baseline prediction (compared to the correct outcome). The Brier Skill Score, BSS (Sutherland et al., 2004a; Sutherland & Obhrai, 2009) was used here. The initial bathymetry was used as the baseline prediction for the final bathymetry, which is common in morphodynamic modelling (Sutherland et al., 2004a).

The BSS is the ratio of the improvement in accuracy of the model over the baseline prediction, compared to the total possible improvement in accuracy. Perfect agreement gives BSS = 1 whereas modelling the baseline condition gives BSS = 0. If the model prediction is further away from the correct bathymetry than the baseline prediction, the skill score is negative. It provides an objective measure of model performance. This skill score is reduced for errors in the position and rate of erosion and deposition and errors in the overall sediment budget.

Composite modelling procedure and example results

A series of numerical and physical model runs were undertaken to test the sensitivity to variations in the set-ups of both models. Neither provides a set of correct results everywhere, as both have weaknesses. Therefore the choice was made to perform a comparative analysis on part of the model domain – in this case an area inshore from the breakwater – where the physical model was judged to give the best results and was taken as the correct bathymetry in the Brier Skill Score. An example set of results, from the numerical model of Test 1 are given in Table 6.3.2.

The Brier Skill Scores in Table 6.3.2 show that including diffraction in the numerical model was more important than including transmission. The skill scores are all relatively high indicating that the numerical model was capable of generating a recirculation current that produced accretion and the formation of a salient, similar to that in the physical model.

Further information about the type of errors involved comes from studying the components of the Brier Skill Score: BSS $= (A_p - B_E - C_{se} + D_{NO})/(1 + D_{NO})$. The position accuracy, A_p, was broadly similar for the 4 numerical model runs of Test Series 1, indicating that the patterns of deposition and erosion were similar. The transport rate

Table 6.3.2 Example skill scores from numerical model of Test 1.

Diffraction	Transmission	BSS	A_p	B_E	C_{SE}	D_{NO}
Yes	Yes	0.58	0.34	0.02	0.02	0.67
Yes	No	0.58	0.32	0.02	0.001	0.67
No	Yes	0.45	0.28	0.13	0.07	0.67
No	No	0.42	0.27	0.15	0.09	0.67

error term, B_E, which should be as low as possible, was much lower for the cases with diffraction than without. In all cases there was a significant normalisation term, D_{NO} which showed that there was a significant change in the measured average bed level, caused by the net import of sediment into the area considered. The sediment budget error, C_{SE}, was a lot lower than the normalisation term, D_{NO}, showing that the numerical model imported approximately the same volume of sand as the physical model.

Conclusions from composite modelling case study

Brier Skill Score gave useful information for optimising information flow

Various physical and numerical model runs were performed with different settings that required different levels of computer, laboratory and human resources to run and which required different information to be passed between models. The skill of each run in simulating the evolution of the salient was calculated, which allowed an optimum model set-up to be chosen that had a high skill score but did not require excessive resources to run.

This procedure forms an objective method for undertaking a sensitivity analysis using the Brier Skill Score. Additional information on the sources of error was obtained using the components of the Brier Skill Score. Using a quantitative measure of skill to assess model sensitivity represents a move away from the use of subjective judgement in choosing a model set-up. However, the objectivity of the method was compromised because there was no truly correct final bathymetry (see below).

Relative results were obtained

The physical model and the numerical model both have strengths and weaknesses, so neither gives the correct bathymetry needed to calculate a Brier Skill Score. A subjective analysis of strengths and weaknesses was undertaken which allowed the authors to choose a limited area of the models where the physical model was judged to give the best results. This model was used as the 'correct' bathymetry and skill scores relative to it were obtained.

Timescales may be different in physical and numerical models

The timescale of the development of the salient may not have been accurately represented in the physical model due to scaling problems (Sutherland & Obhrai, 2009).

A useful extension to the present method would be to take numerical model results at different times and calculate the skill score using the physical model at a set time. If the timescale of bathymetric evolution was not the same in the physical and numerical model, it is likely that the phase accuracy term in the Brier Skill Score will remain similar at different times, but that the amplitude error term, will vary systematically with the length of time that the numerical model was run for. A morphological timescale could then be calculated from the durations of the numerical and physical model runs required to minimise the amplitude error term.

Laser scanner improved transfer of bathymetry from physical model

The use of a commercial laser scanner allowed almost the entire physical model bathymetry to be accurately measured and transferred to the numerical model. Each model was scanned from two directions and about 300,000 points were measured, compared to up to about 1,000 using a mechanical profiler. This meant that the Brier Skill Score could be applied to any area of the model. However, the scans from two directions did not always match well, indicating a shortcoming in its use (see below).

Correct application of procedures is important for quality control

The investigation of the sensitivity of physical model results to the complexity of the boundary conditions was not completed as expected as some of the physical model data was not of sufficient quality. New equipment and methods were applied to this unusual case, but not always successfully. Data quality checks should have been carried out at the time to ensure that the data was of sufficient quality to use before proceeding to the next stage. Procedures for new instruments and methods need to be developed, tested, documented and applied to ensure that high quality data is collected.

Benefits of combining physical and numerical models

A sensitivity analysis could be carried out on the results from a physical model or a numerical model only. The additional benefits of modelling both included:

- An improved understanding of the strengths and weaknesses of the models;
- Physical model results were used in the numerical model setup;
- The numerical model was used to model an entire scheme of offshore detached breakwaters, including the interaction between breakwaters, which could not be done in the physical model because of scaling issues;
- The physical model was used to provide details and limitations on the numerical model results. For example, details of local scour can only be obtained from the physical model. Similarly, the numerical model had no swash zone processes and no undertow in the surfzone, so the physical model gave a better representation of the erosion of the shoreline to either side of the salient.

Physical and numerical models both have roles to play in modelling the morphodynamics responses of beaches to structures. A sensitivity analysis should be carried out as a routine part of any morphological study. This method for using a skill score as an integral part of the sensitivity analysis could be more widely applied. A thorough, case specific, albeit subjective, analysis of the strengths and weaknesses of the models to be used should always be undertaken first, though. It could also be applied to the validation of models (Sutherland et al., 2004a).

The procedure developed allows the incremental benefits of adding additional processes (in a numerical model) or using more complicated boundary conditions (in a physical model) to be assessed. The procedure could be applied to numerical or physical models on their own, or in combination (as here) to aid in the choice of the appropriate model setup. An appropriate set-up should perhaps be the simplest one that will give a suitably skilful prediction, as increasing the number of processes represented does not necessarily improve model skill noticeably and does increase the human and/or computational effort involved in completing a model run.

The central problem in using a skill score with physical and numerical models of bathymetric change is that it is a measure of the accuracy of a prediction (compared to the correct outcome) relative to the accuracy of a baseline prediction (compared to the correct outcome) in situations where there is no model that predicts the correct outcome. Here a subjective analysis of the strengths and weaknesses of the physical and numerical models was used to select an area where there was a best model, which was used as the 'correct' model in the skill score. Hence, relative skill scores were calculated.

The procedure would be improved if both numerical and physical models could be compared to field data so that the model's skill scores could be calculated and strengths and weaknesses identified for generic cases. Some work has already been done for coastal profile models (van Rijn et al., 2003) and for coastal area models (Sutherland et al., 2004b) and a more thorough analysis of models' performance against field data would allow the strengths and weaknesses of different types of models to be quantified. This process would be helped by having more data on the local and far field response of beaches to the installation of structures, which can only come from detailed field monitoring programmes.

This would provide the basis for a more quantified analysis of the strengths and weaknesses of models to be carried out before the selection of the best model and an appropriate area for the sensitivity analysis and would give confidence in the comparative results from the procedure for performing a sensitivity analysis using skill scores developed here.

6.3.6 Composite modelling of scour with incorporation of parameterised turbulence from physical model (DTU)

It has been reported that sediment transport increases markedly with increasing turbulence level (Sumer et al., 2003). Current advanced numerical models do not take into consideration the influence of turbulence on sediment transport and therefore on scour around structures (Sumer, 2007). Recent research (Roulund et al., 2005) indicates that scour is influenced by the effect of externally generated turbulence (turbulence generated by vortex flow processes such as the horseshoe vortex and the vortex shedding).

The objective of the study was to investigate the influence of turbulence on scour using the Composite Modelling (CM) technique. The numerical component of the CM adopted in the study is the same as that used in Roulund *et al.* (2005), where the focus was flow/scour around a circular pile/pier subject to a current flow. As will be discussed later, this model consists of two numerical codes, namely hydrodynamic code (a 3D RANS solver); and morphologic code. The physical-modelling component of the present CM, on the other hand, involved experiments where the influence of turbulence on sediment transport was unveiled by a systematic laboratory investigation.

The CM exercise was done with reference to a half buried spherical object. Our interest on scour around spherical objects comes from scour protection, an area which is of large practical importance in marine civil engineering, considering its application in marine pipelines, offshore wind turbine foundations, breakwaters, etc. One of the key questions is: How does an element in the front row of a stone protection layer respond to the flow? Clearly this is the most exposed part of the protection layer. In the present study, an element of the front row of the protection layer is isolated, and simulated by a spherical body. Observations show that such an individual, spherical element seating on a sediment bed rolls into its scour hole, and is eventually buried with natural backfilling, with the burial depth being S/D = O(0.5) (S refers to equilibrium scour depth, D refers to diameter of the half-buried sphere and O refers to order of magnitude), Truelsen *et al.* (2005). It is on this premise that a **half-buried sphere** is considered in the study. (Incidentally this configuration has also other applications such as sea mines on the ocean bottom (those different from cylindrical mines), and habitat structures installed on river bottoms to offset fish habitat losses.)

Although a substantial amount of knowledge has accumulated on scour around marine and hydraulic structures in recent years, there are no studies investigating the influence of turbulence on scour (Sumer, 2007). The purpose of the study was to obtain a good understanding of the processes related to the influence of externally generated turbulence on scour. This could not be achieved by implementing numerical techniques alone, nor could it be achieved by use of sheer physical modelling techniques. Therefore it was hoped that, by use of composite modelling, the processes involving the externally generated turbulence and its role in scouring could be investigated in a systematic manner. Furthermore, this approach enabled us to switch the externally generated turbulence on and off in the model runs to single out the effect of the latter, and therefore, to shed light onto these complex processes.

Summary of composite modelling technique and results

As mentioned previously, the numerical model (NM) of the CM has two components: (1) Hydrodynamic model, and (2) Morphologic model. The NM is essentially the same as that used in Roulund *et al.* (2005). The hydrodynamic model is a 3D RANS solver, Ellipsys 3D, developed at DTU Fluid Mechanics and RISØ. Turbulence viscosity is calculated by k-omega SST closure model (Roulund *et al.*, 2005) in the present application. The morphologic model, on the other hand, comprises three components: (1) That describing the sediment transport process (Engelund-Fredsøe bedload equation, in vectorial form); (2) That describing the sand slide process (the bed avalanches when the slope exceeds the angle of repose); and (3) That describing the mass balance for sediment (at each grid point on bed, to solve bed elevation h). The morphologic model was modified so that the **influence of turbulence** on morphology was incorporated in the calculations, a feature which was not included in Roulund *et al.* (2005).

Flow around a bottom-seated hemisphere has two key features: Horseshoe vortex in front of the sphere, and lee wake behind it, with shed, arch vortices. The flow underneath these areas exhibits strong turbulence, as illustrated in Figure 6.18. The figure shows Turbulent Kinetic Energy (TKE), k, for two stages of the scour process, calculated from the present NM. The top diagram illustrates the initial stage of the scour process (the bed is plane) and the bottom diagram illustrates the equilibrium stage of the scour process (the scoured bed). The figure indicates that the sphere introduces an additional field of turbulence. The implication of the latter is that the sediment transport (and therefore the scour) will not be same as that described by the classic sediment transport formulae. Hence the sediment transport needs to be modified to

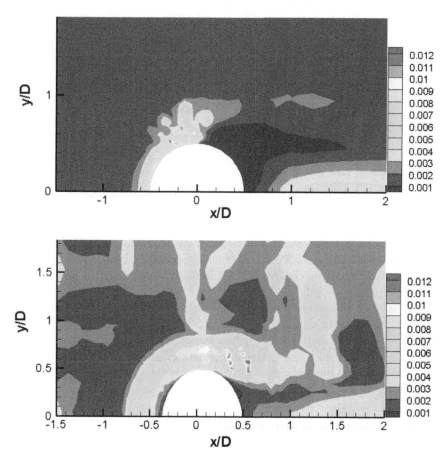

Figure 6.18 Turbulent Kinetic Energy (TKE), **k**. Top: Initial stage of scour, plane bed; and Bottom: Equilibrium stage, scoured bed.

account for the additional turbulence. This is essentially what was done in the present CM study. The modelling procedure is as follows:

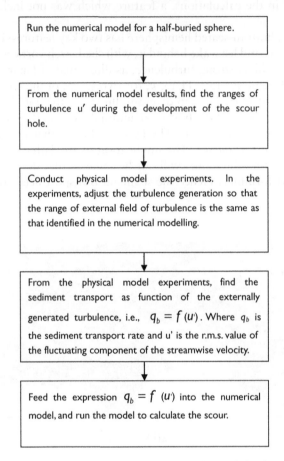

The results of the present CM exercise show that while the equilibrium scour depth is not influenced by the effect of externally generated turbulence, the time scale of scour is affected quite markedly. The equilibrium scour depth is not influenced because the plan-view extents of the horseshoe and lee-wake vortices are the same regardless of whether the externally generated turbulence in the calculations is On or Off (The equilibrium scour depth is presumably $S/D = 0.45$ where S is the equilibrium scour depth and D is the diameter of the half-buried sphere). Our findings show that the time scale of scour with the externally generated turbulence On is a factor of 2 smaller than that with the turbulence Off. This is expected as the sediment transport occurs faster due to the presence of 'additional' turbulence.

Conclusions from the composite modelling case study

We managed to transfer information from the PM to the NM, and as a result, the scour calculation in the NM was significantly. The latter makes it possible to run both PMs and NMs at the same time to obtain a more accurate picture of the scour process, which was not possible before the present CM exercise. In the case of the high Shields-parameter regime, the effect of turbulence on sediment transport may be an issue, which is worth exploring further, although it is expected that the overall effect on the equilibrium scour depth is not very significant (Sumer & Fredsøe, 2002). The results of the present case study are summarized in a paper by Dixen *et al.* (2011).

The present case study may act as a guideline to implement the present technique to other case studies where the involvement of turbulence is profoundly significant in the development of the bed morphology. These cases may include scour around other types of marine and hydraulic structures (bridge piers, piles, offshore wind turbine foundations, breakwaters, pipelines, etc.), or development of bed forms (ripples and dunes) in an otherwise undisturbed flow situations.

The numerical-modelling component of a composite model may not necessarily be the same as when it is applied as a stand-alone numerical model (e.g., the numerical model of the present composite model required much smaller time steps than when applied as a stand-alone numerical model). Likewise, the physical modelling component of a composite model may not necessarily be the same as that when applied as a stand-alone physical modelling.

6.3.7 Parameterisation of physical model results for composite modelling (UHANN)

In this study a large-scale physical model was implemented to understand the hydrodynamic processes around a vertical slender monopile in waves and dimensions of the developing scour in non-cohesive sediment. A review of the current literature was conducted to determine the existing tools used in composite modelling (numerical modelling, physical modelling, analytical modelling and field measurements). Extensive reviews on the topic are given by Whitehouse (1998), Sumer and Fredsøe (1999) and Sumer *et al.* (2001). Constitutive studies on scour development were carried out, for instance by Sumer *et al.* (1992), Kobayashi (1994), Carreiras *et al.* (2000), Sumer and Fredsøe (2001) and Rudolph and Bos (2006). Furthermore, numerical models were developed, which are described in Roulund (2000), Roulund *et al.* (2005), Umeda *et al.* (2006, 2008), Sumer (2007) and Göthel (2008).

The review highlighted that data from field measurements are not available and there are no suitable numerical models for the simulation of wave induced scour around monopiles. Furthermore, some results from the performed large-scale tests were used to validate a numerical model recently developed at a partner-institute (Institute of Fluid Dynamics (Göthel, 2008)) of the Coastal Research Centre at UHANN, but the results were not satisfactory for irregular waves (example of the comparison see Figure 6.19). Therefore this model could not be used for the composite modelling.

The only applicable tools for the CM study were the scour formula of Sumer and Fredsøe (2001, see Figure 6.20) and physical small – and large-scale models. The composite model as defined by Oumeraci (1999, 2009), was not suitable for this study. Subsequently, a case was defined to verify and modify the parameterisations in the analytical model which were derived from physical small-scale tests. The results from large-scale laboratory tests in the Large Wave Channel GWK were used to solve this parameterisation problem. The alternative Composite Modelling procedure to improve the knowledge about scour around vertical slender monopiles is

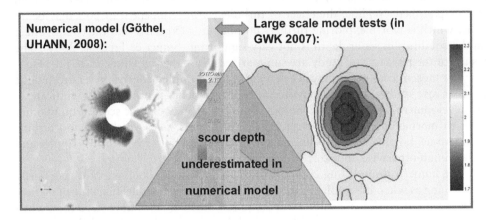

Figure 6.19 Comparison of results derived from a numerical model and physical large-scale tests.

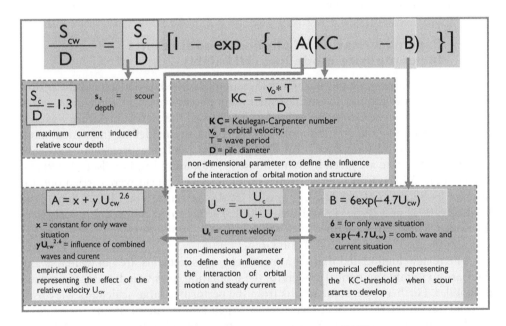

Figure 6.20 Formula of sumer and Fredsøe (2001).

schematised in Figure 6.21. The aim of the conducted composite model was to find a reliable possibility to predict the scour development around a slender vertical mono-pile induced by waves combined with steady tidal currents.

Summary of composite modelling techniques and results

After reviewing the existing analytical approaches, the model derived by Sumer and Fredsøe (2001) was used for the comparative study. This model predicts the equilibrium

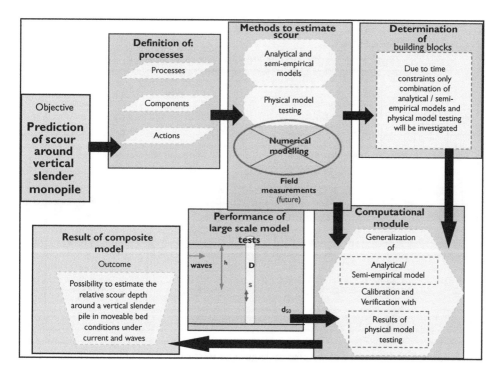

Figure 6.21 The applied composite modelling procedure.

scour depth induced by combined waves and steady currents. The main parameter in this model is the Keulegan Carpenter number (*KC* number), which is determined by the horizontal orbital velocity v_o, the wave period T and the pile diameter D. For wave spectra, Sumer and Fredsøe (2001) recommended the use of the root mean square velocity v_{rms} and the peak wave period T_p. The model was verified and calibrated by small-scale tests. Therefore, scale effects may occur which can reduce the reliability of the results predicted by this model. This leads to the main goal of the Composite Modelling study: to check the parameterisation of the analytical model using results from large-scale laboratory tests. The procedure of the large-scale tests performed within the Composite Modelling study is shown in Figure 6.22.

The measured scour depths of the large-scale model tests were compared to the scour depths calculated with the analytical model using the measured velocities and wave periods. The scour depths observed in the first test sequence with the initial even bottom provided a qualitatively good agreement with the trend given by the analytical model. However, the scour depths measured in the large-scale tests show significant differences. Using the recommended parameters from the small-scale model tests the analytical model underestimates the scour depths.

To find a better agreement of the measured and calculated scour depths the parameter *KC* has been determined using different characteristic velocities and wave periods. The *KC* number varied for the different parameter combinations. After calculating the scour depths using diverse *KC* numbers the depths were compared to

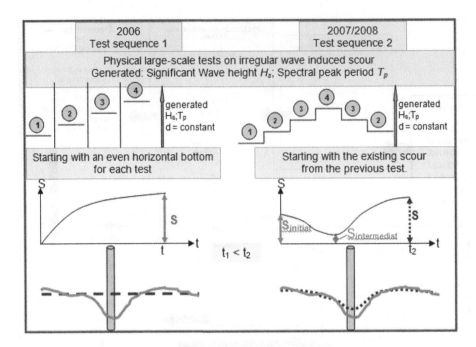

Figure 6.22 Procedure of the physical large-scale tests carried out in the GWK.

the measured scour depths. The agreement of the measured and the calculated scour depths were rated with the sum of the absolute deviations. The best fitting parameter combination was found using the maximum wave velocity v_{max} and the mean wave period T_m to compensate the differences of the scour depths.

The second test sequence was conducted with existing scour in the initial sea bed conditions prior to running tests. Such conditions are similar to field conditions though have never been tested before. The time history of scour development with an existing scour hole differ considerably from that for common tests starting with an even horizontal bottom. Three phases were observed: (1) the scour is partially filled (additional sand transport into the scour from the vicinity of the monopile prevails); (2) the scour hole deepens (scour processes prevail) until (3) the quasi-equilibrium of the scour development is achieved (balanced sand transport in the vicinity and in the scour hole). The ratio of the shear stresses in the vicinity and in the scour hole is assumed to be different to that for an even bottom before starting the test. However, an exact explanation for this type of scour development must be determined by further experimentation.

The tests of the second test sequence were carried out with varying wave conditions simulating the development of an entire storm surge event. Four differing wave spectra were generated as shown in Figure 6.22 in a successive row, firstly with increasing wave energy and afterwards with decreasing wave energy. The evolution of the scour depths with increased wave energy was different from the scour depths with decreased wave energy. For the former, the scour processes proceed faster and the final scour depth is deeper. The filling up of the scour hole had a shorter duration for tests

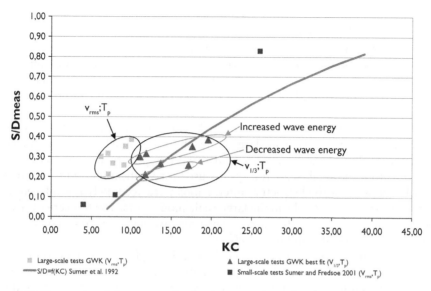

Figure 6.23 Measured relative scour depths (test sequence 2) versus calculated KC number (different parameter combinations).

with increased wave energy compared to those with decreased wave energy. In addition to these findings, a similar comparative analysis of the measured and calculated scour depths was conducted. The results from the large-scale tests show significant differences compared to those derived from the analytical model with respect to trend and order of magnitude (see Figure 6.23 GWK rectangles).

Besides the revision of the *KC* number, the coefficients A and B of the aforementioned analytical formula were modified to find a better agreement of calculated and measured relative scour depths. The coefficients for scour only induced by waves (Sumer *et al.*, 1992) are given with A = 0.03 and B = 6. The scour depths of the first test sequence fit very well with the given coefficients of Sumer and Fredsøe (2001). However, the measured scour depths of the second test sequence fit better with calculated scour depths using coefficients A – 0.025 and B = 7.5. Thus, the initial conditions, namely existing scour hole or even bottom or increased or decreased wave energy, play a decisive role.

Conclusions from the composite modelling case study

A Composite Model as initially planned for this case could not be implemented due to the lack of reliable numerical models and time limitation to develop new models within the frame of the project. Instead, an alternative Composite Model procedure was implemented on the basis of physical large-scale model tests to check and modify the parameterisation of the most commonly used scour formula as derived by Sumer and Fredsøe (2001).

The alternative composite modelling procedure was successful. The comparative analysis of the parameters from the scour formula and those obtained from large-scale model tests allowed the determination of the uncertainties of parameterisation. With further analysis, a parameter combination was found and used to calculate the scour depth. These complemented the measured data from large-scale model tests. A further phase of large-scale tests improved understanding the timing of scour development on the initial seabed conditions (existing scour or horizontal bed) and the considered phase of the storm (increasing or decreasing wave energy). The tests provided a good foundation for further developments and improvements of numerical models.

The composite modelling was conducted with comparative analysis using an analytical model and results from large-scale physical modelling. The parameterisation of the analytical model was examined with respect to possible scale effects which result from the small-scale test results used for verification and calibration of the model. The wave-seabed-structure interactions were accurately reproduced, both qualitatively and quantitatively, in the large-scale model tests. The test series provided reliable data for validation of the models. The previous small-scale tests could be analysed with respect to scale- and model-effects.

With the results of the large-scale model tests, the analytical model was improved to reliably predict scour depths. The improved prediction capability was achieved by using different characteristic parameters for the wave period and the wave induced flow velocity. It was found that the scour evolution depended significantly upon the initial bed conditions. The time history of scour development with an existing scour hole at the beginning of the tests differ considerably from that of tests starting with an even bottom (see Summary of technique and results).

Combined wave and current represent the common forcing conditions in the field. The effect of the different angles between wave and current directions is of practical importance. There were no available laboratory facilities or numerical models to study this effect. The development of a process-orientated numerical model may improve the possibility of predicting reliable scour depths under combined wave and current conditions considerably.

6.3.8　Use of a numerical model to overcome 2D physical model constraints (UPC)

UPC applied the composite modelling to permeable Low Crested Structures (LCS) both submerged and emerged. In particular, the objective of the study was the integration of physical and numerical models to take advantage of the strengths of each and consider the limitations. This was carried out by means of an optimal combination of physical modelling in the UPC CIEM large scale flume (Gironella et al., 1999; Sánchez-Arcilla et al., 2000) and the morphodynamic numerical model LIMORPH (Alsina et al., 2003; Sánchez-Arcilla et al., 2006; Sierra et al., 2009a), that modelled waves, currents and beach morphodynamics. For a complete description of both models see Sierra et al. (2007b, 2008).

The performed tests (Sierra et al., 2008; 2009b) involved obtaining (both in the PM and the NM at scale 1:1) wave heights, currents, water level and bottom evolution around a permeable structure located on a sloping beach (1/15) in a wave flume with mobile bed (sand with mean grain size 200 μm). Water depth at the frontal toe of the structure was 1 m. Two types of structures were employed: one submerged with a freeboard of 0.25 m and one emerged, with a crest height of 0.15 m. Both structures were built with cubic concrete blocks with a porosity of 0.5. The structure slope was of 1V:2H at both sides and the crest width 1.5 m. Nine irregular wave conditions (generated with a Jonswap spectrum) were tested for each structure, corresponding to three different wave heights and three different wave steepnesses.

Although other possible approaches were analyzed (Sierra *et al.*, 2007a, 2007b, 2008), the composite modelling was carried out using a selection of models to give better performance. The assessment of the areas of better performance was process-based, i.e. taking into account the physical processes that each model could simulate accurately in each area (in front or behind the structure). Thus, a better representation of hydro-morphodynamic conditions around a permeable structure was obtained as the results given by the composite model were obtained by a model that, *a priori*, could simulate all the main physical processes taking place in this area. On the contrary, each individual model (PM or NM) could not simulate all the involved processes in the whole domain (in front and behind the structure). In particular some limitations and constraints observed in 2DV physical models were overcome with the employed approach.

The proposed composite modelling approach was based on the fact that NM and PM had varying performances at different areas of the studied domain. Thus, 2DV physical models worked well at the structure front because they reproduced shoaling, refraction, reflection and breaking processes. However, they also had some limitations. For example, they did not work well at the lee side of the structure. Although they reproduced wave transmission, they did not simulate diffraction, longshore currents and sediment transport. Furthermore, they and were limited in reproducing the water levels, due to wave set-up (piling-up) behind the breakwater. This gave rise to spurious results for this variable.

The numerical model did not work well at the structure front. Although it reproduced shoaling, refraction and breaking, it did not simulate reflection. At the lee side of the structure, it measured diffraction effects, which could not be included in the PM. Moreover, it gave more realistic set-up results at the leeside of the structure, because it did not suffer from water pile-up. However, this model type did not accurately perform wave transmission.

Therefore, both types of modelling have advantages and could be complementary. For this reason, they were combined in a way that the results of the model that performed better in an area of the studied domain were selected for the modelling process. In particular, PM results were selected in the front of the structure and NM results at the leeside of the structure, expanding also the domain from 2DV to 2DH or Q3D. This approach was applied to permeable low crested structures (both emerged and submerged). Thus, as indicated in the previous section, a better representation of hydro-morphodynamic conditions around a permeable structure was obtained

Summary of composite modelling technique and results

The modelling approach adopted in this study is illustrated in Figure 6.24:
   The study involved the following steps:

- The wave model capabilities were improved to take into account wave transmission through and over the structure. These improvements were calibrated with physical model results;

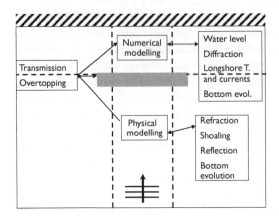

Figure 6.24 Composite modelling approach.

- Two geometries for permeable LCS were defined, one corresponding to emerged structures and the other to submerged ones;
- Representative offshore wave conditions (in the Spanish Mediterranean Sea) were defined and used by both the numerical and the physical model;
- The initial physical model was set up. The domain of the physical model included the LCS and the shoreline;
- The computations of the numerical model were set up on two different domains: one equal to that of the physical model (including the LCS) and the other domain similar but extended to both lateral sides;
- A number of physical experiments (9 for each type of structure) were carried out. In these experiments the bottom evolution was measured as well as several hydrodynamic parameters (wave heights, current velocities, water levels, etc.);
- The numerical model was run in the same domain and with the same parameters that the physical model, without activating the diffraction terms, in order to calibrate the wave heights with the results of the physical model. In this step there was interaction between both models;
- The calibrated numerical model was run with the same parameters but on the extended domain and taking into account diffraction effects;
- The results of each model were combined, selecting the areas where they gave better performances for permeable low crested structures: the physical model at the structure front and the numerical model at the lee side of the structure.

Figure 6.25 shows an example of the bottom evolution results obtained applying the proposed composite modelling approach.

The study resulted in a gaining of experience and expertise in the application of composite modelling. In particular, knowledge was gained on how to correct features that cannot be simulated by the numerical model (e.g. transmission through the structure) and are not well reproduced by the physical model (e.g. diffraction effects and water pilling up at the leeside of the structure). The determination of the bottom

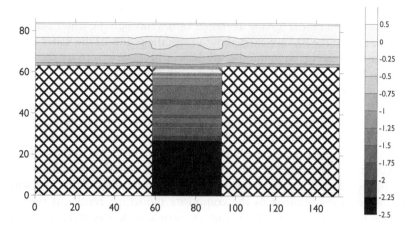

Figure 6.25 Composite modelling for emerged structures. Final bathymetry after 9 tests.

evolution around a given permeable LCS geometry in the chosen physical scale model facility was more efficient and better represented. The expansion of the application domain from 2DV to 2DH/Q3D depended on the models used.

Conclusions from the composite modelling case study

The study of problems in coastal zones is very complex. In particular, in cases such as permeable structures, there are many processes involved and some of these are very difficult to reproduce. PM or NM alone cannot cover all the processes and do not give a similar output for different parts of the studied domain. The combination of both model types allows increasing the number of processes reproduced (although this does not include all the involved processes) and the extension of the domain with a better description of hydro-morphodynamic processes.

Both PM and NM have strengths and weaknesses that have been highlighted during this work. The combination of both model types can overcome some of the weaknesses and enhance the whole modelling capability and the quality of the results, but is not enough to reproduce all the processes in the whole domain. It was almost impossible to run both models simultaneously and in real time, since they have different temporal scales and, as a consequence, they take different periods of time for their execution. For this reason, one of the possible approaches (to run both models in an iterative way) had to be discarded.

One of the weaknesses of the model was that the 2DV PMs underpredicted wave fields at the leeside of the structure as they did not take diffraction effects into account. On the other hand, they overpredicted water levels in this area due to the pilling-up of water. As a consequence, it was difficult to calibrate water levels in the NM at the leeside of the structure due to this pilling-up effect. Difficulty was experienced in evaluating errors in the employed instruments, in particular in wave gauges and current meters.

In this study the modelling of currents through the porous structure was problematic. Currents pose further problems such as difficulties in calibrating the NM. The NM is a vertically integrated model (giving averaged values through the water column) whilst the measurements in the PM are carried out in specific points of the water column. For this reason, as well as the large number of devices necessary to cover the whole area, the current field cannot be defined accurately with the PM in front of the structure.

The model also suffered from lack of data in some areas. In the selected approach, PM covers the front of the structure and NM the leeside that can be extended laterally to take into account diffraction and longshore effects. The lateral areas located seawards of the structure are not covered by the PM because there, the wave conditions are different (there is not reflection). If the NM results were considered in these lateral areas, then a discontinuity between PM data and NM results would appear. This is the main limitation of the selected approach for the composite modelling of this case study.

There were discontinuities in the composite modelling results at the transition area between both types of models (i.e. on the structure shoreward end), because the results of both models are different. Nevertheless, this is a minor problem for some parameters. For beach morphodynamics, the structure itself poses a discontinuity since in its location bottom evolution is not computed. For other magnitudes like wave height and water level, this discontinuity between the results of both models actually exists, but due to the different processes being present at both sides of the structure (wave diffraction, water pilling-up, etc.), these discontinuities can be assumed as part of the perturbations introduced by the presence of the structure.

Transferring information between the two types of models proved difficult. The diffraction effects and bottom evolution were not calibrated in the NM as the former could not be reproduced by the PM and the latter involved different processes in both models (wave diffraction, longshore sediment transport, etc.). Therefore, the results relative to this process are only evaluated from a qualitative point of view. Furthermore, the lower morphodynamic resolution of NM did not allow the reproduction of some of the features (e.g. ripples). Finally, in spite of extending the domain from 2DV to 2DH, only wave normal incidence can be simulated.

With respect to the qualitative and quantitative measures used, the PM had a higher spatial resolution for morphodyamics and bottom evolution than the NM. Conversely, the NM had a higher spatial resolution for hydrodynamic process than the PM and the results are easier to obtain. Hydrodynamic measurements in the PM is expensive both in cost (it requires to deploy a large number of instruments) and in time (to process the time series recorded).

Finally, it must be stressed that although the main objectives of this composite modelling exercise were achieved (extension of the domain from 2DV to 2DH/Q3D and overcoming of most of the constraints and limitations that the single models have) a number of uncertainties still persist as well as some processes that cannot be reproduced by this technique, as indicated above.

The modelling approach used in this study can be used for similar case studies looking at hydro-morphodynamic processes around permeable structures parallel to the shoreline. However, there are some considerations that need to be taken into account. This approach is useful when only a 2DV physical model (wave flume) is used, since it allows extending the domain to 2DH/Q3D, reproducing more processes and in a more accurate way and overcoming some of the constraints imposed by the use of a 2DV model. If a 2DH/Q3D physical model (wave basin) is used, the approach is meaningless.

One of the limitations of this composite modelling technique was that it could only be applied to problems with waves having a normal incidence to the structure. The technique can also be applied to obtain the bottom evolution, at the leeside of the structure, only when there is no interest in studying small features (e.g. ripples).

Due to the discontinuity problem, the technique can only be applied to cases where the interest is focused on the bottom evolution in the area located between the structure and the shoreline and/or to the study of scouring problems at the structure toe. It cannot be applied when modelling the whole 2DH domain. In order to improve the accuracy of the results given by this approach, the simulation of water and sediment fluxes through the structure should be included in the NM. In the same way, the use of Q3D or 3D NM instead of 2DH numerical models, would allow a better modelling of the currents.

6.4 ISSUES IN APPLYING COMPOSITE MODELLING

6.4.1 Reflections on composite modelling during CoMIBBS

Numerical model – requiring explicit description of the processes – scales of interest

There is an increasing number of situations that can be satisfactorily modelled using a numerical model. This is particularly so for situations where one process is considered, and the range of temporal and spatial scales of interest is limited. For these, well-tested explicit mathematical equations are available, and have been implemented in NM codes, that mimic the dynamic processes of interest, given the "drivers and constraints" (geometry, bathymetry, forcing, etc.) that we provide. Examples are: wave modelling in a coastal zone, wave modelling in a harbour, tidal modelling, spreading of pollutants in the near-coast. For these, the scales of interest are also the scales modelled. When the interest is in much larger scales than modelled by the models, the situation becomes more complicated. A well known example is using shallow water flow models to derive mean flows or mean currents. For non-linear processes, the intrinsic interaction across scales when considering larger scales is not necessarily well represented by the commonly used shallow water flow models – one should therefore be careful to interpret its results for the larger scales. One should also be aware of this when using the current output from such short term models to drive transport models, if interest is in larger spatial and temporal scales.

One common use of NMs was in the design of the PMs, where NMs were used to help decide the physical model scale, bathymetry, location of the structure being modelling within the facility, plus the locations of the wave paddles and wave guides, and input conditions.

Physical models: Natural representation of processes – similarity principle – scale effects

Other situations, particularly those involving non-linear interactions between water, structures and sediment can best be modelled using physical models or PM. This is due to the similarity principle which holds for physical models – "the same natural processes occur – essentially we just scale them" (Kamphuis, 2000; Van Os *et al.*, 2004).

The interactions do not have to be explicitly formulated – nature automatically reproduces them, given the drivers and constraints provided. In these cases the physical model should be run at a large enough scale to minimise scale effects. In some cases, such as with a detached breakwater, it is not possible to run at a scale large enough to have no scale effects and these must be accepted as a limitation of the physical model. The case of one of the CoMIBBS partners – to analyse optimal scales of a PM, is a very relevant one.

The physical model can be used:

- To obtain **site-specific results** from complicated flow / sediment / structure interactions that cannot be modelled using a bottom-up process-based NM, due to limitations in the representation of processes in sets of equations that mimic (that is: describe sufficiently accurately for our purposes) the dynamics of the processes on our scales of interest.
- To provide **systematic information on a physical process** that can be mathematically formulated or parameterised and incorporated in a NM code (i.e. the standard method of numerical model development).

Coupling of models – exchange of quantified information

In our composite modelling, this raises questions such as "how can the results of the site-specific model be incorporated into a NM of a wider area?" For example, in one of the cases of a detached breakwater studied in CoMIBBS, only a single breakwater could be included in the PM (to minimise scale effects) while the entire scheme consisted of four breakwaters. What information from the PM is needed by the NM, and how can we formulate or quantify this? Transfer of information between PM and NM has to be quantified – simply because the PM and the NM are forced by explicit, quantified information on water level, wave height and period, current, concentration, etc.

Identifiability and measurability of processes

Quantified information is relatively easily obtained for quantities that are well identified, with a large signal to noise ratio. Examples are wave heights, wave periods, water level – heights and timing (phase), currents: speed, direction and timing (phase). A key characteristic is that in most coastal situations, these quantities are strongly varying, due to the dynamics of their processes. The same holds true for water depth. While temperature, salinity and concentrations of dissolved matter are often much less dynamic, local point measurements are highly identifiable. The situation is more complex for concentrations of substances with a patchy character (e.g. suspended sediment) and if interest is in larger time scales than the typical model time scales. Examples are spatial distribution of salinity and temperature, of concentration of sediment, erosion and deposition patterns and their evolution in time. While we may be able to determine local values sufficiently accurately, how accurate is the difference of the value measured between the end and the beginning of an experiment? If the local measurement error is of the order of 2%, errors in changes are easily a full order higher. This is relevant when considering erosion and deposition or sediment redistribution, which is the spatial distribution of changes in bathymetry. In several of the CoMIBBS experiments the quantification of sediment volume and distribution, and change in bathymetry were assessed as this information needed to be transferred between PM and NM or vice versa.

Role of prior analysis in reducing dimensionality

In two cases, the success of a data-fitting scheme depended on prior analysis to identify the key non-dimensional parameters controlling the problem. This reduced the dimensionality of the problem, which reduced the computational effort in calculating a best-fit to the data.

Importance of bathymetry

When cases of sediment transport and morphological changes are studied, the information that comes from a PM essentially includes bathymetry. It is often possible to run a series of PM tests, to analyse the resulting bathymetries and to parameterise the results in some way. For example, in another of the CoMIBBS cases studied, the scour depth at a monopile subject to wave action was characterised in terms of its Keulegan-Carpenter number. This form of empirical parameterisation leads to an analytical or behavioural model that directly links the response to the forcing, without trying to represent the processes involved (although knowledge of the processes is useful in guiding the choice of the parameters used in the fitting exercise).

Coastal area NMs, however, solve the equations of fluid motion and calculate sediment concentrations, which are then advected and diffused, leading to sediment erosion and deposition through the conservation of mass. Although every coastal morphological model is behavioural at some space and time scale, these models incorporate far more of the processes than the analytical models derived from fitting curves to physical model bathymetry data. It is therefore difficult to incorporate the bathymetric results from a physical model of a small area into a coastal area numerical model of a large area.

In a further CoMIBBS case, an NM was set up for the whole domain, plus a PM for that part of the domain where NM results of the interactions with the structures are known to be much less reliable than those of the PM. The bathymetry from the PM is copied into the NM. However, due to differences in the representation of the processes, this may lead to discontinuities in surface elevation or wave quantities and changes in the overall sediment budget, which needed to be assessed.

Use of PM to calibrate model parameters – skill scores

A PM can be used to calibrate or validate a NM. The NM process descriptions are verified and if needed, the differences are interpreted to guide adjustments in NM model descriptions and parameter settings. Model parameters are tuned until the model gives the closest response to the physical model. All this requires explicit choices on which are the relevant quantities that need to be evaluated and compared, and how we will quantify their differences (and which differences in which quantities we find acceptable). This can be done in a quantified manner using a skill score. Skill scores, such as the Brier Skill Scores are defined to immediately show whether resultants of a model simulation are or are not improved over those of a reference simulation.

Skill scores for optimising PM and NM model set up

A variation on this is to use skill scores to choose an optimum PM or NM set-up that has a high skill score but does not require excessive resources to run. However, even a calibrated NM cannot be expected to reproduce all the features of the PM as it does not include all the processes involved. For example, scour holes due to wave reflections and shoreline changes driven by swash zone sediment transport cannot be generated in most coastal area NMs as the formulations describing the relevant processes are not included.

Bathymetry variation and transfer between PM and NM

One important principle that could be used to reconcile the bathymetries from a physical and a numerical model is the conservation of sediment. In the CoMIBBS case study of the beach evolution inshore from a detached breakwater, the statistics calculated included a sediment budget (or map-mean) error term that is the relative difference between the changes in seabed elevation from the PM and the NM. Alternatively the difference between mean bed levels at the end of the PM and the NM runs could be used. When either of these error terms is zero the amount of sediment imported into or out of the relevant area will have been the same in the PM and the NM. In this case, the NM of a system of breakwaters will include in the far field effects the correct import or loss of sediment from behind a single breakwater.

There will still be differences between the final bathymetries from the PM and NMs. The approach to dealing with them may depend on the use to which the results will be put. A designer may choose to take a conservative approach and use the lower bathymetry at each point, which would lead to an apparent net loss of sediment. Alternatively both bathymetries may be presented along with the changes between them. The changes give an indication of the potential errors that come from using the NM versus the PM. If the PM was free from scale effects and measurement errors the changes would give a direct measure of the errors that come from using the NM.

If a series of corresponding PM and NM tests are run, it may be possible to parameterise the differences so that intermediate cases may be interpolated to. This would be done as part of the post-processing of the data. In one of the CoMIBBS cases this was extended further into modelling these differences.

In the case where a scour hole is generated in a PM, but not in an NM, it may be possible to calculate the volume of sediment lost from the scour hole and include this in the NM as a point source of sediment (in the way that sediment dispersion from a dredger is modelled). However, during the formation of, for example, a scour hole caused by waves plunging over the shoulder of a breakwater, the eroded sand is often deposited locally, so a sediment dispersion approach may not be appropriate.

Role of data assimilation techniques to minimise differences between NM and PM

It may be possible to use a data assimilation scheme to improve, for example, the numerical modelling of currents inshore of a detached breakwater, using flow data from a PM and this may improve the numerical modelling of bathymetry evolution. This approach will not be able to reproduce processes that are not in the model, such as scour or swash-zone sediment transport.

Last but not least, we note that physical and numerical models are usually run by separate teams of people, sometimes even from separate institutes. It is important that modellers understand the role their model is playing in the Composite Modelling approach. Information exchange between modellers, as well as models, should be encouraged.

6.4.2   Good modelling practice

Modelling plays a role in the process that starts with a question or set of questions, and that subsequently should lead to an answer or set of answers that is adequate for the originally posed question. The answer sometimes leads to a second set of questions. This process, from the initial question to the final answer and its interpretation in terms of fitness for use, is often called the Modelling Cycle. In recent decades, the complexity of the Modelling Cycle has increased. On the one hand, for reasons that the models are more complex, and sometimes consist of chains of models, on the other hand, because different persons, with different backgrounds, each play a role in this cycle. It may start with parliament that wishes to have options evaluated to minimise (recreational) beach erosion and evaluate cost of mitigating measures. The responsible cabinet minister and her staff assess the question within the departmental policy framework and pass on this question to specialists, who do their interpretation, and suggest possible approaches to analyse and answer this problem. Scientists, engineers, computer specialists are often be involved in the subsequent steps. The results of the analysis, forming the basis for the conclusions and answer to the originally posed question, travel the same way back. Along the road, data, perceptions and interpretations are communicated to guide the process. Understandably, miscommunication may easily occur.

The above issues were addressed explicitly some years ago in a project called Good Modelling Practice (van Waveren et al., 1999). The objective of the project was to make the Modelling Cycle explicit through formulating questions, answers and interpretations. Often some standardised format is used, in order not to forget issues, and to promote re-use and referencing. This way the process and its decision points are made transferable and referable to all involved. It can be easily discussed and adjusted, and the process can be backtracked. The EU project Harmoniqua in the early 2000s extended this further by introducing computer based tools to facilitate this process, given much attention to user friendliness, re-use and automated build-up of experiences and using this to improve the system. For further information on Good Modelling practice see for instance Refsgaard and Henriksen (2004), Old et al. (2005), Refsgaard et al. (2006), Scholten et al. (2007), and the Harmoniqua website (www.harmoniqua.org).

Note: A topic directly related to the systematic process of model application outlined as Good Modelling Practice is the systematic analysis and validation of the models that are used. A IAHR initiative to devise a hydraulic community approach how to do this and how to subsequently document the validity of the models is described in Dee *et al.* (1994).

Below some key questions in Good Modelling Practice are highlighted, in this case, for the somewhat simplified situation that the researcher or engineer himself formulates the task, and more or less starts from scratch. It involves addressing the relevant questions of problem analysis and potential approaches to the modelling and problem solving:

- What are the key processes of interest that need to be represented?
- What (ranges of) length scales and spatial scales need to be resolved for these processes?
- What is the area or domain of the case of interest?
- What do we know from literature and other earlier studies about this area and these processes?
- Can the questions that one wishes to answer be derived from measurable quantities?
- Can they be answered if one knows the quantities that are determined by our models (for example: water level, currents, concentrations, wave energy density); c.q. can they be easily and uniquely derived from those quantities?
- What level of accuracy and reliability is required: will a trend type answer or other simplified approach be sufficient?
- What field data sets that describe the processes for this particular case are available?
- What are the uncertainties associated with these data?
- Are the data sufficient to answer the modelling questions by means of a data analysis and desk study?
- Can the problem be suitably modelled by physical experiments (scales, laboratory effects)?
- Do I have access to such a facility to model this particular problem?
- What are its strengths and weaknesses?
- Are consistent field data sets or other supporting information available for adequately forcing a physical scale model, and for properly and quantitatively assessing its results?
- Are quantified process descriptions available – which set of equations applies?
- Are these appropriate for the spatial and temporal scales?
- Can these be solved or approximated by simple linear models?
- Do I have computer codes available to me that are fit to simulate the problem?
- Which are their strengths and weaknesses?
- Are they stable, accurate, robust and well documented?
- Are consistent field data sets or other supporting information available for adequately forcing a numerical model, and for properly and quantitatively assessing its results?
- In short: what is the state of the art, which models/tools are available and suited for the problem?

These steps are first steps for any problem analysis. The answers to these questions will determine which of element or elements of the methodology presented in Figure 6.1 will be best suited for further analysis of the problem at hand. Essentially, this implies a strengths and weaknesses analysis of the various elements and options, before a choice is made.

At the end of this analysis – so, before any actual modelling has taken place -, one should also have made an assessment on aspects such as:

- Sufficiency and suitability of the data for forcing the model, including associated uncertainties.
- Sufficiency and suitability of the data for quantified assessment of modelling quality
- Identifiability of the key quantities that allow the assessment of the relevant processes.
- Suitable techniques for quantification (e.g. skill scores) and visualisation of the quantities and the processes for the various modelling options.

The results of this analysis and considerations should generally be written up, as part of the Good Modelling Procedures. Only then one can make an objective, well-founded choice on the modelling approach. The advantage of doing this in the above way is that all considerations and choices are written down explicitly. They are now transferable and reproducible, and can be used in discussions with colleagues and clients, as part of the quality assurance process. An analysis like the above was also made by the CoMIBBS partners, as their first activity in the project. It was used to guide the detailed definition of the modelling experiments (Gerritsen *et al.*, 2007).

6.4.3   Selection of composite modelling – hypothesis

It is essential that a problem analysis, in the form of the question and answers that were discussed in the previous Section, be carried out in order to estimate *a priori* the potential benefits of using Composite Modelling –

"the integrated and balanced use of physical and numerical models".

The state of the art of the tools for analysing the problem needs to be critically determined first. In different words and somewhat aggregated:

- Which individual approaches and models are available?
- Which level of sophistications is required / which model or set of models?
- Are their strengths and weakness such that they can be satisfactorily applied to analyse the problem given the type of answer required?

The answers to these questions will or should give the information whether it makes sense to use composite modelling:

- Is one type of model not adequate for analysing and solving the problems?
- Do physical and numerical models complement each other for this problem?
- Can different scales be separated in time and / or in space?

- Can the interfacing of the models and the exchange of data be defined uniquely, in terms of quantified parameters?
- Can an estimate be given of the uncertainty of the transferred data, so the transfer and effect of uncertainties from model to model can be estimated?

In this respect, see the approaches in e.g. Oumeraci (1999), Kamphuis (2000) and Van Os *et al.* (2004). They all point out the importance of this interfacing, which needs to be quantified, taking into account uncertainties involved. The above provides the material to formulate a hypothesis on what we expect to gain by composite modelling for this problem, for example in terms of:

- Increased cost effectiveness;
- Improved accuracy;
- Reducing uncertainties;
- Tackling problems that could not reasonably be solved by a single model approach.

The outcome of this process is an argued decision on whether or not composite modelling is needed, or whether it is beneficial for modelling the problem at hand. The advantage of approaching this conscientiously lies in the fact that we systematically build up the various elements into, for instance, an exchangeable and easily accessible library of physical and numerical models and their descriptions or metadata, that is, updates on their capabilities, usefulness, and track records. This is part of Good Modelling Practice.

6.4.4   Setting up a composite modelling experiment

The results of section 6.4.2 and 6.4.3 can be used to decide whether we should set up a composite modelling experiment or not. In order to do so successfully, we have to decide:

- What models do we use?
- What is the overall domain in space and time for the problem?
- What time and space scales will be covered by each model?
- How to make external forcing consistent for PM and NM?
- What data to transfer between the models, to provide forcing for the other model?
- How these are best quantified?
- What is the uncertainty in the data and how do these uncertainties influence the other model?
- How do we quantify the results – in terms of which quantities and in which norms?
- How do we evaluate PM and NM data against field data ?
- How do we visualise
- How can a sensitivity analysis be conducted (this includes a data exchange protocol)?

A rigorous data exchange protocol is advised in order to enable the assessment of the influence of the partial modelling elements on each other. Similarly, quantification and evaluation of the results is important in order to show the additional gain or benefit of applying composite modelling.

By systematically addressing these issues and reporting the selections and arguments behind it, we make the problem analysis process transparent, and therefore transferable. This is not only important for the particular project itself. Sharing with colleagues internally and externally, and learning from each other will benefit from this as well.

6.4.5   Pitfalls and traps

In principle, applying a Good Modelling Practice approach such as discussed and outlined in the previous sections of this chapter should lead to a well-defined problem approach. It includes well-founded decisions on the use of models or model elements, and *a-priori* assessment of what we expect to gain by applying composite modelling. Still, it may be helpful to the modeller to be aware of conditions and constraints that may reduce and limit the anticipated success of composite modelling. Such conditions and constraints are not uncommon for any traditional modelling project either, and they also occurred during CoMIBBS.

First of all, composite modelling is not the medication against all evils, or the solution for all problems. Many problems can adequately be addressed by either PM or NM. So, one should be critical in the assessment of what one formulates as expected benefits by applying composite modelling. Secondly, combining or balanced use of physical models with numerical models is a fairly new field. It is not simpler than traditional modelling – on the contrary, it is rather more difficult. That is certainly true now, when the hydraulic modelling community still has to build up its knowledge and experience with it. We do it for a certain case, because we have analysed that it will have clearly expected benefits over traditional modelling approach for the case at hand. So, at this stage, in most cases composite modelling is not cheaper but more expensive than traditional modelling. However, when we decided to do composite modelling, we had concluded that it would be beneficial in a well defined sense – not being able to solve the problem with traditional approaches, reduction of uncertainties, etc. We may have to offset these benefits against the additional cost.

A last point is a suitable balance between optimism and realism. Modellers tend to be quite optimistic about what they can achieve. That is a good quality, as it is the basis on which progress is made. In addressing composite modelling, however, a good measure of realism is necessary next to optimism. It is not unknown to define a problem with a scope that is just too complex, due to range of scales, interaction of processes, measurability of results, etc. If a case study is conceived too broadly, and not sufficiently precisely defined, it will be extremely difficult to make any modelling successful. This is of course not unique to Composite Modelling, but in this new field, the risk of this happening is larger.

"Management of expectations" – of the modellers in this case, is a key issue.

References

Aage, C. & Sand, S.E. (1984). Design and Construction of the DHI 3-D Wave basin. *Proc. Symp. Description and Modelling of Directional Seas*, Technical University of Denmark.

Aberle, J., Koll, K. & Dittrich, A. (2008). Form induced stresses over rough gravel-beds, *Acta Geophysica*, 56, 584–600.

Abernethy, R. & Olliver, G. (2002). Effects of modelling long and short crested seas on overtopping of a vertical faced breakwater under normal wave attack, *Paper 190 in International Conference on Coastal Engineering*, abstracts publn. Thomas Telford, London.

Ahrens, (1989). Stability of reef breakwaters. *Journal of Waterway, Port, Coastal and Ocean Engineering*, American Society of Civil Engineers, 115(2), 221–234.

Alben, S., Shelly, M. & Zhang, J. (2002). Drag reduction through self similar bending of a flexible body. *Nature*, 420, 479–481.

Alsina, J.M., Cáceres, I, Sánchez-Arcilla, A., González-Marco, D., Sierra, J.P. & Montota, F. (2003). Morphodynamics in the neighbourhood of a submerged breakwater. *Proc. 3rd IAHR Symposium on River, Coastal and Estuarine Morphodynamics*, Barcelona, Spain, 1018–1028.

Andersen, O.H. & Fredsøe, J. (1983). Transport of Suspended Sediment along the Coast. Inst. Of Hydrodyn. & Hydr. Eng., Techn. Univ. of Denmark, *Progr. Rep.* 59, 33–46.

Andersen, K.H. & Faraci, C. (2003). The wave plus current flow over vortex ripples at an arbitrary angle. *Coastal Engineering*, 47(4), 431–441.

Arnskov, M.M., Fredsøe, J. & Sumer, B.M. (1993). Bed shear stress measurements over a smooth bed in three-dimensional wave-current motion. *Coastal Engineering*, 20, 277–316.

Badiei, P., Kamphuis, J.W. & Hamilton, D.G. (1994). Physical experiments on the effects of groins on shore morphology. *Proc. 24th International Conference on Coastal Engineering*, ASCE, 99, 1782–1796.

Baglio, S., Faraci, C., Foti, E. & Musumeci, R. (2001). Measurements of the 3D scour process around a pile in an oscillating flow though a stereo vision approach. *Measurement*, 30, 145–160.

Baldock, T.E. & Simmonds, D.J. (1999). Separation of incident and reflected waves over sloping bathymetry. *Coastal Engineering*, 38, 167–176.

Barmuta, L.A., McKenny, C.E.A. & Swain, R. (2001). The response of lotic mayfly *Nousia* sp. (Ephemeroptera: Leptophlebiidae) to moving water and light of different wavelengths. *Freshwater Biology*, 46, 567–573.

Battisto, G.M., Friedrichs, C.T., Miller, H.C. & Resio, D.T. (1999). Response of OBS to mixed grain size suspensions during Sandy Duck'97. *Coastal Sediment Conference 99*, ASCE, New York. pp. 297–312.

Battjes, J.A., Bakkenes, H.J., Janssen, T.T. & Van Dongeren, A.R. (2004). Shoaling of subharmonic gravity waves. *Journal of Geophysical Research*, 109.

Bejan, A. (2005). The constructal law of organization in nature: tree-shaped flow and body size. *The Journal of Experimental Biology*, 208, 1677–1686.

Bendat, J. & Piersol, A. (1971). *Random Data: Analysis and Measurement Procedures*. Wiley, New York.

Benoit, M. (1993). Practical comparative performance survey of methods used for estimating directional wave spectra from heave-pitch-roll data. *Proc. 23rd Int. Conf. Coastal Engineering*, Vol. 1, ASCE, 62–75.

Benoit, M. & Teisson, C. (1994). Laboratory Comparison of Directional Waves Measurement Systems and Analysis Techniques. *Proc 24th Int. Conf. Coastal Engineering, ASCE*, Japan, paper 255.

Benoit, M., Frigaard, P. & Schäffer, H.A. (1997). Analysing multidirectional wave spectra: A tentative classification of available methods. *Proc. IAHR Seminar Multi-directional Waves and their Interaction with Structures*. 27th IAHR Congress, San Francisco, Aug 10–15, 28 pages.

Bettess, R. (1990). Survey of lightweight sediments for use in mobile-bed physical models. *Movable Bed Physical Models*, Kluwer Academic Publishers, The Netherlands, pp. 115–123.

Bezuijen, A., Wolters, G. & Müller, G. (2005). Failure mechanisms for blockwork breakwaters. *Proc. ICE Coastlines, Structures and Breakwaters Conf. 2005*, ICE, London.

Biésel F. (1951). Les appareils generateurs de houle en Laboratoire. *La Huille Blanche*, 6(2,4 and 5).

Biésel, F. & Suquet, F. (1951). *Les appareils générateurs de houle en laboratoire*, La Houille Blanche, No. 4. English translation: Laboratory wave-generating apparatus, project report No. 39, St. Anthony Falls Hydraulic Laboratory, Univ. of Minnesota, 1954.

Biewener, A.A. (Ed.) (1992). *Biomechanics Structures and Systems: a practical approach*. Oxford University Press.

Birkemeier, W.A. (1985). Field Data on Seaward Limit of Profile Change. *J. Waterways, Harbours & Coastal Eng. Div.*, 111(3), 598–602.

Boller, M.L. & Carrington, E. (2007). Interspecific comparison of hydrodynamic performance and structural properties among intertidal macroalgae. *Journal of Experimental Biology*, 210(11), 1874–1884.

Borgman, L.E. Directional Spectra Models for Design. *Proceedings of the 1st Annual Offshore Technology Conference (No. 1069)*, Houston, Texas, Houston.

Bouma, T.J., De Vries, M.B., Low, E., Peralta, G., Tánczos, I.C., Van De Koppel, J. & Herman, P.M.J. (2005). Trade-offs related to ecosystem engineering: A case study on stiffness of emerging macrophytes. *Ecology*, 86(8), 2187–2199.

Brasington, J. & Smart, R.M. (2003). Close range photogrammetric monitoring of experimental drainage basin evolution. *Earth Surface Processes and Landforms*, 28(3), 231–247.

British Standards Institution (1991). *British Standard Code of practice for Maritime structures, Part 7 (BS 6349), Guide to the design and construction of breakwaters*, British Standards Institution, London.

Broderick, L. & Ahrens, J.P. (1982). *Rip-rap stability scale effects, Technical Paper 82–3*, U.S. Army Engineer Waterways Experiment Station, Vicksburg, MS.

Brossard, C., Delouis, A., Galichon, P., Granboulan, J. & Monadier, P. (1990). Navigability in channels subject to siltation: physical scale model experiments. *Proc. 22nd Int. Conf. Coastal Eng.*, ASCE, New York, 3088–3101.

Bruun, P. (1954). Coast Erosion and the Development of Beach Profiles. *Tech. Memo. 44*, Beach Erosion Board, Corps of Engineers.

Bullock, G., Obhrai, C., Wolters, G., Müller, G., Peregrine, H. & Bredmose H. (2004). Characteristics and design implications of breaking wave impacts. *Proc. 29th Int. Conf. Coastal Engineering, Lisbon*, 3966–3978.

Bullock, G., Obhrai, C., Müller, G., Wolters, G., Peregrine, H. & Bredmose, H. (2005). Advances in the understanding of wave impact forces. *Proc. ICE Coastlines, Structures and Breakwaters Conf. 2005*, ICE, London.

Bullock, G.N., Obhrai, C., Peregrine, D.H. & Bredmose, H. (2007). Violent breaking wave impacts. Part 1: Results from large-scale regular wave tests on vertical and sloping walls. *Coastal Engineering*, 54, 602–617.

Buckingham, E. (1914). On physically similar systems: illustrations of the use of dimensional equations. *Physical review*, 4(6), 345-3-376.

Burchard, H., Craig, P.D., Gemmrich, J.R., van Haren, H., Mathieu, P-P., Meier, H.E.M., Smith, W.A.N., Prandke, H., Rippeth, T.P, Skyllingstad, E.D., Smyth, W.D., Welsh, D.J.S. & Wijesekera H.W. (2008). Observational and numerical modeling methods for quantifying coastal ocean turbulence and mixing. *Progress in Oceanography*, 76, 399–442.

Burcharth, H.F. & Liu, Z. (1992). Design of Dolos Armour Units. *Proceedings of the 23rd International Coastal Engineering Conference*, American Society of Civil Engineers, 1, 1053–1066.

Caillat, J.M. (1987). Physical scale models and siltation in estuaries. *Proc. 2nd Int. Conf. of Port Engng. in Developing Countries (COPEDEC)*, Nanjing Hydraulic Research Institute, China Ocean Press, 1321–1331.

Campbell, L.J., McEwan, I., Nikora, V., Pokrajac, D., Gallagher, M. & Manes, C. (2005). Bed-load effects on hydrodynamics of rough-bed open-channel flows. *Journal of Hydraulic Engineering*, 131, 576–585.

Carreiras, J., Larroudé, P., Seabra-Santos, F. & Mory, M. (2000). Wave scour around piles. *Proceeding of the International Conference on Coastal Engineering*, Sydney, 1860–1871.

CEM Coastal Engineering Manual. (2006). '*Part VI; Introduction to coastal project element design*', Document nr. EM 1110-2-1100.

CERC, (1984). *Shore Protection Manual, Vol. I–III*, Coastal Engineering Research Centre, Vicksburg.

Chadwick, A., Ilic, S. & Helm-Petersen, J. (2000). An evaluation of directional analysis techniques for multidirectional, partially reflected waves, Part 2: application to field data. *Journal of Hydraulic Research*, 38(4), 253–258.

Chandler, J.H., Wackrow, R., Sun, X., Shiono, K. & Rameshwaren, P. (2008). Measuring a dynamic and flooding river surface by close range digital photogrammetry. *The International Archives of the Photogrammetry, Remote Sensing and Spatial Information Sciences*, CHEN Jun, JIANG Jie, Ammatzia PELED, International Society for Photogrammetry and Remote Sensing, xxi Congress, Beijing, 3rd July, 211–216.

CIRIA, CUR & CETMEF (2007). *The Rock Manual. The use of rock in hydraulic engineering* (2nd edition). C683, CIRIA, London.

Clauss, G.F. (2002). Dramas of the sea: episodic waves and their impact on offshore structures. *Applied Ocean Research*, 24, 147–161.

Cohen de Lara, M. (1955). Coefficient de perte de charge en milieu poreux base sur l'equilibre hydrodynamique d'um massif. *La Houille Blanche*, No. 2.

Cooper, G.G., Callaghan, F.M., Nikora, V.I., Lamouroux, N., Statzner, B. & Sagnes, P. (2007). Effects of flume characteristics on the assessment of drag on flexible macrophytes and a rigid cylinder. New Zealand Journal of Marine and Freshwater Research, 41(1), 129–135.

Cornelisen, C.D. & Thomas, F.I.M. (2006). Water flow enhances ammonium and nitrate uptake in a seagrass community. *Marine Ecology Progress Series*, 312, 1–13.

D'Amours, O. & Scheibling (2007). Effect of wave exposure on morphology, attachment strength and survival of the invasive green alga Codium fragile ssp. tomentosoides. *Journal of Experimental Marine Biology and Ecology*, 351, 129–142.

Daemrich, K-F. (1999). Overtopping at Vertical Structures. *Second German-Chinese Joint Seminar on Recent Developments in Coastal Engineering*, Tainan, Taiwan, Republic of China, September 13th to 15th 1999, Coastal Ocean Monitoring Center (COMC).

Dai, Y.B. & Kamel, A.M. (1969). *Scale effect tests for rubble-mound breakwaters.*, U.S. Army Engineer Waterways Experiment Station, Corps of engineers, December 1969, Vicksburg, Mississippi.

Dalrymple, R.A. (1985). *Physical modelling in coastal engineering*, A.A. Balkema, Rotterdam.

Davidson, M.A., Kingston, K.S. & Huntley, D.A. (2000). New Solution for Directional Wave Analysis in Reflective Wave Fields. *Journal of Waterway, Port, Coastal and Ocean Engineering*, 126(4), 173–181.

Davis, W.P. (1970). Closed systems and the rearing of fish larvae. *Helgoländer wiss. Meeresunters*, 20, 691–696.

De Lemos, M.J.S. (2006). *Turbulence in porous media: Modeling and applications*, Elsevier Ltd, Oxford, 335 p.

De Rouck, J. & Geeraerts, J. (2005). *CLASH Final Report D46*, University of Gent.

Dean, R.G. (1985). Physical modelling of littoral processes. **In:** Dalrymple, R.A. (Ed) *Physical modelling in coastal engineering*, Rotterdam.

Dean, R.G. & Dalrymple, R.A. (1991). Water Wave Mechanics for Engineers and Scientists. *Advanced Series on Ocean Engineering, Vol. 2*, World Scientific Press.

DHI (2007). *LITPACK. Noncohesive Sediment Transport in Currents and Waves. User Guide.* Danish Hydraulic Institute, Denmark.

Dibajnia, M. & Watanabe, A. (1996). A transport rate formula for mixed-size sands. *Proc. 25th Int. Conf. Coastal Eng.*, Orlando, Florida, ASCE, 3971–3803.

Dibajnia, M. & Watanabe, A. (1998). Transport rate under irregular sheet flow conditions. *Coastal Engineering*, 35, 167–183.

Dick, D., Cunge, J., Labadie, G., Mateo, A.R., Mathiesen, M. Price, R., Santos, M. & Warren, R. (1994). *Guidelines for documenting the validity of computational modelling software.* IAHR bulletin, 24 p.

Dietrich, W.E, Kirchner, J.W., Ikeda, H. & Iseya, F. (1989). Sediment supply and the development of the coarse surface layer in gravel-bedded rivers. *Nature*, 340, 215–217.

Dixen, M., Hatipoglu, F., Sumer, B.M. & Fredsøe, J. (2008). Wave boundary layer over a stone-covered bed. *Coastal Engineering*, 55, 1–20.

Dixen, M., Sumer, B.M. & Fredsøe, J. (2011). *Flow and scour around a half-buried sphere.* In preparation.

Ducrot, V., Cognat, C., Mons, R., Mouthon, J. & Garric, J. (2006). Development of rearing and testig protocols for a new freshwater sediment test species: The gastropod *Valvata piscinalis*. *Chemosphere*, 62, 1272–1281.

Eggert, W., Daemrich, K-F. & Kohlhase, S. (1982). Two and Three-Dimensional Investigations on Permeable Breakwaters Including Irregular Waves. *Proc. 3. Intern. Conf. on Water Resources Development*, Bandung.

Einstein, H.A. (1950). The bed-load function for sediment transportation in open channel flows. *Techn. Bulletin 1026*, US Dept of Agriculture.

Elgar, S. & Guza, R.T. (1985). Observations of bispectra of shoaling surface gravity waves. *Journal of Fluid Mechanics*, 161, 425–448.

Elgar, S. & Chandran, V. (1993). Higher-order Spectral Analysis to Detect Nonlinear Interactions in Measured Time Series and an Application to Chua's Circuit. *International Journal of Bifurcation and Chaos*, 3, 19–34.

Elgar, S., Herbers, T.H.C., Chandran, V. & Guza, R.T. (1995). Higher-order Spectral Analysis of Non-linear Ocean Surface Gravity Waves. *Journal of Geophysical Research*, 100(C3), 4977–4983.

Environment Agency. 2003. *River Habitat Survey in Britain and Ireland: Field Survey Guidance Manual.* River Habitat Survey Manual: 2003 version, Environment Agency, 136 p.

Evensen, G. (1994). Sequential data assimilation with a nonlinear quasi-geostrophic model using Monte Carlo methods to forecast error statistics. *Journal of Geophysical Research*, 99, 143–162.

Faraci, C., Foti, E. & Baglio, S. (2000). Measurements of sandy bed scour processes in an oscillating flow by using structured light. *Measurement*, 28, 159–174.

Faraci, C. & Foti, E. (2002). Geometry, migration and evolution of small scale bedforms generated by regular and irregular waves. *Coastal Engineering*, 47, 35–52.

Faraci, C. & Foti, E. (2006). Statistical sensitivity of roughness in presence of ripples. *Proc. 30th International Conference on Coastal Engineering*, ASCE, Vol. 3, 2428–2240.

Fathi-Maghadam, M. & Kouwen, N. (1997). Nonrigid, nonsubmerged, vegetative roughness on floodplains, *Journal of Hydraulic Engineering*, 123(1), 51–57.

Finelli, C.M., Hart, D.D. & Merz, R.A. (2002). Stream insects as passive suspension feeders: Effects of velocity and food concentration on feeding performance. *Oecologia*, 131(1), 145–153.

Folkard, A.M. (2005). Hydrodynamics of model Posidonia oceanica patches in shallow water. *Limnology and Oceanography*, 50(5), 1592–1600.

Fonseca, M.S., Koehl, M.A.R. & Kopp, B.S. (2007). Biomechanical factors contributing to self-organization in seagrass landscapes. *Journal of Experimental Marine Biology and Ecology*, 340(2), 227–246.

Foti, E. & Blondeaux, P. (1995). Sea ripple formation: the heterogeneous sediment case. *Coastal Engineering*, 25, 237–253.

Foti, E., Marini, A. & Musumeci, R.E. (2007). Structured light approaches for measuring sandy bottom evolution. *HYDRALAB-III, SANDS Meeting*, Budapest, 19 Nov 2007.

Franco, C., Van der Meer, J.W. & Franco, L. (1996). Multi-directional Wave Loads on Vertical Breakwaters. *Coastal Engineering 1996, Conference Proceedings, American Society of Civil Engineers*, Orlando, USA, 2008–2021.

Frechette, M., Butman, C.A. & Geyer, W.R. (1989). The importance of boundary-layer flows in supplying phytoplankton to the benthic suspension feeder, *Mytilus edulis L. Limnology & Oceanography*, 34(1), 19–36.

Friedrichs, M. & Graf, G. (2009). Characteristic flow patterns generated by macrozoobenthic structures. *Journal of Marine Systems*, 75(3–4), 348–359.

Frigaard, P. & Brorsen, M. (1995). A time domain method for separating incident and reflected irregular waves. *Coastal Engineering*, 24, 205–215.

Gacia, E., Granata, T.C. & Duarte, C.M. (1999). An approach to measurement of particl flux and sediment retention within seagrass (Posidonia oceanic) meadows. *Aquatic Botany*, 65, 255–268.

Gambi, M.C., Nowell, A.R.M. & Jumars, P.A. (1990). Flume observations on flow dynamics in *Zostera marina L.* (eelgrass) beds. *Marine Ecology Progress Series*, 81, 159–169.

Galbraith, R.V., MacIsaac, E.A., Macdonald, J.S. & Farrell, A.P. (2006). The effect of suspended sediment on fertilization success in sockeye (*Oncorhynchus nerka*) and coho (*Oncorhynchus kisutch*) salmon. *Canadian Journal of Fisheries and Aquatic Sciences*, 63(11), 2487–2494.

Gallagher, E.L., Elgar, S. & Thornton, E.B. (1998). Megaripple migration in a natural surf zone. *Nature*, 394(9), 165–168.

Galland, J.C. (1994). Rubble-mound Breakwaters stability under Oblique Waves: an Experimental Study. *Proc 24th International Conference on Coastal Engineering*, ASCE, paper 156.

Garcia-Hermosa, I., Knaapen, M., Obhrai, C. & Sutherland, J. (2009). *CoMIBBS numerical modelling report*. HR Wallingford Report TR175, Release 2.0, March 2009. Also HYDRALAB-III Report JRA1-09-0.

Gauthier, J.P. & Kupka, I.A.K. (1994). Observability and Observers for Nonlinear Systems. *SIAM J. Control Optim.*, 32, 975–994.

Gerritsen, H., Caires, S. & Van den Boogaard, H.F.P. (2007). Interim report on analysis of functionality of models. *HYDRALAB-III Deliverable JRA1.1* for Consortium Participants and the EC.

Ghisalberti, M. & Nepf, H. (2002) Mixing layers and coherent structures in vegetated aquatic flow. *J. Geophys. Res.*, **107(C2)**, 1–11.

Ghisalberti, M. & Nepf, H. (2006). The structure of the shear layer over rigid and flexible canopies. *Environmental Fluid Mechanics*, **6**, 277–301.

Gironella, X. & Sánchez-Arcilla, A. (1999). Hydrodynamic behaviour of submerged breakwaters. Some remarks based on experimental results, *Proc. Coastal Structures'99*, Santander, Spain, 891–896.

Goda, Y. (1970). A Synthesis of Breaker Indices. *Trans. Japan Soc. Civil Eng.* **2(2)**, 227–230.

Goda, Y. & Suzuki, Y. (1976). Estimation of Incident and Reflected Waves in Random Wave Experiments. *Proceedings of the 15th Coastal Engineering Conference*, American Society of Civil Engineers, **1**, 828–845.

Goda, Y. (1985). Random Seas and Design of Maritime Structures. Tokyo: Uni. Tokyo Press.

Goda, Y. (1986). Effect of wave tilting on zero-crossing wave heights and periods. *Coastal Engineering in Japan*, **29**, 79–90.

Goda, Y. (2000). *Random seas and design of maritime structures, Advanced Series on Ocean Engineering*, Vol. 15, World scientific, Singapore.

Gosselin, F., Langre, E. & Machado-Almeida, B.A. (2009). Drag reduction of flexible plates by reconfiguration. *Submitted to Journal Fluid Mechanics (In Review)*.

Göthel, O. & Zielke, W. (2006). Numerical modelling of scour at offshore wind turbines. *Proceeding of the Coastal Engineering Conference 2006*, 2343–2353, San Diego, California.

Göthel, O. (2008). Numerical modelling of flow- and wave- induced scour around vertical circular piles. *PhD-Thesis Institute of fluid dynamics, Leibniz University of Hannover*, Report-No. 76/2008.

Graham, G.W. & Manning, A.J. (2007). Floc size and settling velocity within a Spartina anglica canopy. Continental shelf Research, **27**, 1060–1079.

Günther, B. & Morgado, E. (2003). Dimensional analysis revisited. *Biological Research*, **36(3–4)**.

Guoren, D. (1998). Keynote Lecture: Development of physical model studies on sediment transport in China. *Proc. 7th Int. Symp. River Sedimentation*. Hong Kong, China (Balkema, Rotterdam, ISBN 90 5809 034 5).

Gurnell, A.M., van Oosterhout, M.P., de Vlieger, B. & Goodson, J.M. (2006). Reach-scale interactions between aquatic plants and physical habitat: River Frome, Dorset. River *Research and Applications*, **22**, 667–680.

Gurnell, A.M., O'Hare, J.M., O'Hare, M.T., Dunbar, M.J. & Scarlett, P.M. (2010). An exploration of associations between assemblages of aquatic plant morphotypes and channel geomorphological properties within British rivers. *Geomorphology*, **116(1–2)**, 135–144.

Guza, R.T. Thornton, E.B. & Holman, R.A. (1984). Swash on Steep and Shallow Beaches. *Proc 19th International Conference on Coastal Engineering*, ASCE, 708–723.

Hallermeier, R.J. (1981). A profile zonation for seasonal sand beaches from wave climate. *Coastal Eng.* 4(3), 253–277.

Hamm, L. & Migniot, C. (1994). Elements of cohesive sediment: deposition, consolidation and erosion. In: *Coastal Estuarial and Harbour Engineers' Reference Book*, Abbott, M.B., Price, N.A. & Spon, F.N. (Eds), London, United Kingdom.

Hamm, L. & Peronnard, C. (1997). Wave parameters in the nearshore: a clarification. *Coastal Engineering*, **32**, 119–135.

Hashimoto, N. & Kobune, K. (1988). Directional Spectrum Estimation from a Baysian Approach. *Proc. 21st International Conference on Coastal Engineering*, ASCE, pp. 62–76.

Hashimoto, N., Nagai, T. & Asai, T. (1993). Modification of the extended maximum entropy principle for estimating directional spectrum in incident and reflected wave field. *Rept. Of P.H.R.I.* 32(4), 25–47.

Hasselmann, K., Munk, W. & MacDonald, G. (1963). Bispectra of Ocean Waves. **In:** Rosenblatt, M. (Ed). Time Series Analysis, New York: John Wiley.

Hentschel, B.T. & Larson, A.A. (2005). Growth rates of interface-feeding polychaetes: Combined effects of flow speed and suspended food concentration. *Marine Ecology Progress Series*, **293**, 119–129.

Herbers, T.H.C. & Burton, M.C. (1997). Nonlinear shoaling of directionally spread waves on a beach. *Journal of Geophysical Research*, **102(C9)**, 21, 101–21, 114.

Hirano, H. (1971). River bed degradation with armouring. *Trans. Japan. Society of Civil Engineers*, **195**, 55–65.

Hiraishi, T. (1997). Wave Directionality to wave action on coastal structures. **In:** Mansard, E. (ed.) *IAHR Seminar – Multi-directional Waves and their Interaction with Structures*. XXVII IAHR Congress, San Francisco.

Hirayama, T. (1997). Modelling of Multi-directional Waves in Naval Architectural Field. **In:** Mansard, E. (ed.) *IAHR Seminar – Multi-directional Waves and their Interaction with Structures*. XXVII IAHR Congress, San Francisco.

Hudson, R.Y. (1959). Laboratory Investigation of Rubble-Mound Breakwaters. *Journal of the Waterways and Harbors Division*, American Society of Civil Engineers, **85(WW3)**, 93–121.

Hudson, R.Y. & Davidson, D.D. (1975). Reliability of Rubble-Mound Breakwater Stability Models. *2nd Symposium on Modelling Techniques*, American Society of Civil Engineers, Vol. 2, 1603–1622.

Hudson, R.Y., Herrmann, F.A., Sager, R.A., Whalin, R.W., Keulegan, G.H., Chatham, C.E. & Hales, L.Z. (1979). *Coastal Hydraulic Models*, Special report No. 5, US Army Engineer Waterways Experiment Station, Vicksburg, Mississippi.

Huettel, M. & Gust, G. (1992). Impact of bioroughness on interfacial solute exchange in permeable sediments. *Mar. Ecol. Prog. Ser.*, **89(2–3)**, 253–267.

Hughes, S.A. & Fowler, J.E. (1990). Validation of movable-bed modeling guidance. *Proc. 22nd Int. Conf. Coastal Eng.*, Delft, The Netherlands, 2457–2470.

Hughes, S.A. (1993a). Physical models and laboratory techniques in coastal engineering. *Advanced Series on Ocean Engineering, Vol. 7*, World Scientific Publishing, Singapore.

Hughes, S.A. (1993b). Laboratory wave reflection analysis using co-located gauges. *Coastal Engineering*, **20**, 223–247.

Hulscher, S.J.M.H. (1996). *Formation and migration of large-scale, rhythmic sea-bed patterns: a stability approach*. Ph.D. Thesis, University of Twente.

Hurther, D. & Lemmin, U. (2008). Improved turbulence profiling with field adapted acoustic. Doppler velocimeters using a bi-frequency Doppler noise suppression method. *J. Atmos. Oceanic Technol.*, **25(2)**, 452–463.

Hwang, M. & Seinfeld, J.H. (1972). Observability of nonlinear systems. *Journal of Optimization Theory and Applications*, **10**, 67–77.

HYDRALAB (2004). Strategy paper 'The future role of experimental methods in European hydraulic research – towards a balanced methodology'. *Journal of Hydraulic Research*, **42(4)**, 341–356.

HYDRALAB, (2007). *Guideline document: "Wave research"*, Task group NA 3.1, Krikegaard, J. (Ed.), EC contract no. 022441 (RII3).

Ilic, S., Chadwick, A. & Helm-Petersen, J. (2000). An evaluation of directional analysis techniques for multidirectional, partially reflected waves, Part 1: numerical investigations. *Journal of Hydraulic Research*, **38(4)**, 243–251.

Isaacson, M. (1991). Measurement of Regular Wave Reflection. *Journal of Waterway, Port, Coastal and Ocean Engineering*, ASCE, **117(6)**, 553–569.

Isobe, M. & Kondo, K. (1984). Method for Estimating Directional Wave Spectrum in Incident and Reflected Wave Field. *Proc. 19th International Conference on Coastal Engineering*, 467–483.

Isobe, M. (1990). Estimation of Directional Spectrum Expressed in Standard Form. *Proc 22nd International Conference on Coastal Engineering*, ASCE, 647–660.

ISU Fisheries Extension (2009). Managing Iowa Fisheries: Aquatic Plant Management. Iowa State University.

Ito, M. & Tsuchiya, Y. (1984). Scale-model relationship of beach profile. *Proc. 19th Int. Conf. Coastal Eng.*, Houston, USA, **Vol. 2**, 1386–1402.

Ito, M. & Tsuchiya, Y. (1986). Time scale for modeling beach change. *Proc. 20th Int. Conf. Coastal Eng.*, Taipei, Taiwan, **Vol. 2**, 1196–1200.

Ito, M. & Tsuchiya, Y. (1988). Reproduction model of beach change by storm waves. *Proc. 21st Int. Conf. Coastal Eng.*, Malaga, Spain, **Vol. 2**, 1544–1557.

Ito, M., Murakami, H. & Ito, T. (1995). Reproducibility of beach profiles and sand ripples by huge waves. *Coastal Dynamics*, pp. 698–708.

James C.S., Birkhead A.L., Jordanova, A.A. & O'Sullivan J.J. (2004). Flow resistance of emergent vegetation. *Journal of Hydraulic Research*, **42**(4), 390–398.

James, S.C., Grace, M.D., Ahlmann, M., Jones, C.A. & Roberts J.D. (2008). Recent Advances in Sediment Transport Modelling. *World Environmental and Water Resources Congress*, Ahupua'a.

Jamieson, W.W., Mogridge, G.R. & Brabrook, M.G. (1989). Side absorbers for laboratory wave tanks, *Proc. 23rd IAHR congress, Ottawa*, C-135–C-142.

Janssen, P., Van Bendegom, L., Van den Berg, J., De Vries, M. & Zanen, A. (1979). *Principles of River Engineering*. Pitman, London.

Järvelä, J. (2002). Flow resistance of flexible and stiff vegetation: a flume study with natural plants. *Journal of Hydrology*, **269**, 44–54.

Jazwinsky, A. (1970). Stochastic Processes and Filtering Theory. Academic Press.

Jensen, O.J. & Klinting, P. (1983). Evaluation of scale effects in hydraulic models by analysis of laminar and turbulent flows. *Coastal Engineering*, **7**(4), 319–329.

Jonsson, I.G. (1963). Measurements in the turbulent wave boundary layer. *Proc. 10th Congr. IAHR*, London, **Vol. 1**, 85–92.

Jonsson, I.G. & Carlsen, N.A. (1976). Experimental and theoretical investigations in an oscillatory rough turbulent boundary layer. *J. Hydr. Res.*, **14**(1), 45–60.

Jonsson, P.R., van Duren, L.A., Amielh, M., Asmus, R., Aspden, R.J., Daunys, D., Friedrichs, M., Friend, P.L., Olivier, F., Pope, N., Precht, E., Sauriau, P-G. & Schaaff, E. (2006). Making water flow: A comparison of the hydrodynamic characteristics of 12 different benthic biological flumes. *Aquatic Ecology*, **40**, 409–438.

Kailath, T. (1980). *Linear Systems*. Prentice Hall: New Jersey.

Kamphuis, J.W. (1972). Scale selection for mobile bed wave models. *Proc. 13th Int. Conf. Coastal Eng.*, Vancouver, Canada, **Vol. 2**, 1173–1195.

Kamphuis, J.W. (1982). Coastal mobile bed modelling from a 1982 perspective. *Research Report No. 76*, Queen's University, Kingston, Ontario, Canada.

Kamphuis J.W. & Kooistra, J. (1990). Three dimensional mobile bed hydraulic model studies of wave breaking, circulation and sediment transport processes. *Proc. Canadian Coastal Conference 1990*, Kingston Ontario. National Research Council of Canada, Associate Committee on Shorelines (ACOS) pp. 363–386.

Kamphuis, J.W. (2000). Designing with models. *Proceedings 27th International Conference on Coastal Engineering*. Edge, B.L. (Ed) ASCE, 19–32.

Karcz, I. (1973). Reflections on the origin of small-scale longitudinal streambed scours. In: Morisawa, M. (Ed.) *Fluvial Geomorphology, The Binghamton Symposia in Geomorphology*, Int. Series No. 4, Allen and Unwin, London, UK, 149–177.

Kemp, P.H. & Simons, R.R. (1982). The interaction between waves and turbulent current: waves propagating with current. *J. Fluid Mech.*, **116**, 227–250.

Kent, T.R. & Stelzer, R.S. (2008). Effects of deposited fine sediment on life history traits of Physa integra snails. *Hydrobiologia*, **596**(1), 329–340.

Kleinhans, M.G. (2005). Upstream sediment input effects on experimental dune trough scour in sediment mixtures. *J. Geophys. Res.*, **110**, F04S06, doi:10.1029/2004 JF000169, 2005.

Klopman, G. & Van Leeuwen, P.J. (1990). An efficient method for the reproduction of non-linear random waves. *International Conference on Coastal Engineering ASCE, Proc*, **Vol. 1**, 478–488, Delft.

Klopman, G. & Van der Meer, J.W. (1998). Random wave measurements in front of reflective structures. *Journal of Waterway, Port, Coastal and Ocean Engineering*, ASCE.

Kobayashi, T. (1994). 3-D Analysis of Flow around a Vertical Cylinder on a Scoured Bed. *Proceedings of the Intern. Conf. on Coastal Engineering*, Venice, Italy, **Vol. 3**, 3482–3495.

Koehl, M.A.R. (1999). Biological biomechanics: life history, mechanical design, and temporal patterns of mechanical stress. *Journal of Experimental Biology*, **202**, 3469–3476.

Koehl, M.A.R. (2003). Physical modelling in biomechanics. *Phil. Trans. R. Soc. London*, **358**, 1589–1596.

Krogstad, H.E. (1988). Max Likelihood Estimation of Ocean Wave Spectra from General Arrays of Wave Gauges. *Modelling, Identification and Control*, **9**, 81–97.

Kuhnle, R.A. (1989). Bed surface size changes in gravel bed channel. *J. Hydraulic Eng.*, **115**, 731–743.

Kumar, M. & Anantha, S. (2007). A numerical and experimental study on tank wall influence on drag estimation. *Ocean Engineering*, **34**(1), 192–205.

Lamberti, G.A. & Steinman, A.D. (1993). Research in Artificial Streams: Applications, Uses, and Abuses. *Journal of the North American Benthological Society*, **12**(4), 313–384.

Langhaar, H.I. (1951). *Dimensional analysis and theory of models*. J. Wiley and Sons, New York.

Langre, E. (2008). Effect of wind in plants. *Annual Reviews Fluid Mechanics*, **40**, 141–169.

Lara, J.L., Garcia, N. & Losada, I.J. (2006). RANS modelling applied to random wave interaction with submerged permeable structures. *Coastal Engineering*, **53**(5–6), 395–417.

Lauder, G.V. & Madden, P.G.A. (2006). Learning from fish kinematics and experimental hydrodynamics for roboticists. *Intern. Journal for Autm. Comput.*, **4**, 325–335.

Le Dimet, F.X. & Talagrand, O. (1986). Variational algorithms for analysis and assimilation of meteorological observations: theoretical aspects. *Tellus*, **38A**, 97–110.

Le Méhauté, B. (1957–58). Perméabilité des diques en enrochments aux ondes de gravité périodiques. *La Houille Blanche*, No. 6 (1957), No. 2 (1958), No. 3 (1958).

Le Méhauté, B. (1976). Similitude in coastal engineering. *Journal of the Waterways, Harbors and Coastal Engineering Division*, American Society of Civil Engineers, **102**(WW3), 317–335.

Le Méhauté, B. (1990). Similitude, in Ocean Engineering Science. **In:** Le Méhauté, B. (Ed.), *The Sea*, New York: John Wiley and Sons. Vol. 9, Part B, 955–980.

Lee, J.K., Roig, L.C., Jenter, H.L. & Visser, H.M. (2004). Drag coefficients for modeling flow through emergent vegetation in the Florida Everglades. *Ecological Engineering*, **22**(4–5), 237–248.

Leonard, L.A., Hine, A.C. & Luther, M.E. (1995). Surficial Sediment Transport and Deposition Processes in a Juncus roemerianus Marsh, West-Central Florida. *Journal of Coastal Research*, **11**(2), 322–336.

LNEC (2008). *COMIBBS Task 2: Composite Modelling Report*. LNEC'S Report. Lisbon.

Long, R.B. & Hasselmann, K. (1979). A Variational Technique for Extracting Directional Spectra from Multicomponent Wave Data. *J. Phys. Oceanography*, **9**, 373–381.

Longuet-Higgins, M.S. & Stewart, R.W. (1964). Radiation stress in water waves: a physical discussion with application. *Deep Sea Res.*, **11**, 529–562.

López, F. & García, M.H. (2001). Mean flow and turbulence structure of open-channel flow through non-emergent vegetation. *Journal of Hydraulic Engineering*, 127, 392–402.

Loudon, C., Best, B.A. & Koehl, M.A.R. (1994). When does motion relative to neighbouring surfaces alter the flow through an array of hairs. *J. Exp. Biology*, 193, 233–254.

Lundgren, H. & Sørensen, T. (1957). A Pulsating Water Tunnel. *Proc. 6th Int. Conf. Coastal Eng.*, Gainesville, Palm Beach and Miami Beach, Florida, pp. 356–358.

Lundgren, H. & Klinting, P. (1987). Rigorous Analysis of Directional Waves. *Proc. of Wave Analysis and Generation in Lab Basins, 22nd Congress IAHR*, 351–363.

Lynett, P. & Liu, P.L.-F. (2002). *COULWAVE Code Manual*, Cornell University Long and Intermediate Wave Modeling Package.

Lyrge, A. & Krogstad, H.E. (1986). Maximum Entropy Estimation of the Directional Distribution in Ocean Wave Spectra. *J. Phys. Oceanography*, 16, 2052–2060.

Manes, C., Pokrajac, D., Coceal, O. & McEwan, I. (2008). On the significance of form-induced stress in rough wall turbulent boundary layers. *Acta Geophysica*, 56, 845–861.

Mansard, E.P.D. & Funke, E.R. (1980). The measurement of incident waves and reflected spectra using a least squares method. *17th Int. Conf. of Coastal Engineering, Sydney*.

Mansard, E.P., Sand, S.E. & Funke, E.R. (1985). Reflection Analysis if Non-linear Regular Waves. *Hydraulics Lab Tech Rep TR-HY-01*, Nat Res Council of Canada, Ottawa.

Mansard, E.P. & Funke, E.R. (1988). Physical Experiments in Laboratories: Contrast of Methodologies and Results. Unpublished lecture notes for the Short Course on Planning and Designing Maritime Structures held prior to the 21st Int. Coastal Engineering Conference in Malaga, Spain.

Mansard, E.P.D. (1991). On the experimental verification of non-linear waves. *International Conference on Coastal and Port Engineering in Developing Countries, COPEDEC III, 20–25, Sep. 1991*, Mombasa, Kenya.

Maoxiang, G. & Zhenquan, K. (1988). An analysis of Second Order Waves in a Wave Tank by 3-probe Correlation Analysis and by Square-Low Operation. *Ocean Eng.*, 15(4), 319–343.

Marsden, R.F. & Juszko, B.A. (1987). An Eigenvector Method for the Calculation of Directional Spectra from Heave, Pitch and Roll Buoy Data. *J. Phys. Oceanography*, 17, 2157–2167.

Marth, R., Müller, G., Klavzar, A., Wolters, G., Allsop, N.W.H. & Bruce T. (2005). Analysis of blockwork coastal structures. *Proc. 29th Int. Conf. Coastal Engineering, Lisbon*, 3878–3890.

Massel, S.R. (1998). Ocean surface waves: their physics and prediction. *Advanced Series on Ocean Engineering*, Vol. 11, World Scientific, London.

Massel S.R. & Brinkman, R.M. (1998). On the determination of directional wave spectra for practical applications. *Applied Ocean Research*, 20(6), 357–374.

MAST-G6M (1993). On the methodology and accuracy of measuring physico-chemical properties to characterize cohesive sediments. *Internal report G6M-Coastal Morphodynamics European project* (EC MAST-I research programme).

McBride, M., Hession, W.C., Rizzo, D.M. & Thompson, D.M. (2007). The influence of riparian vegetation on near-bank turbulence: A flume. *Earth Surface Processes and Landforms*, 32(13), 2019–2037.

McIntire, C.D. (1993). Historical and other perspectives of laboratory stream research. In: Lamberti G.A. & Steinman A.D. (Eds) Research in Artificial Streams: Applications, Uses, and Abuses. *Journal of the North American Benthological Society*, 12(4), 313–384.

McLelland, S.J., Ashworth, P.J., Best, J.L. & Livesey, J.L. (1999). Turbulence and secondary flow over sediment stripes in bimodal bed material. *J. Hydraulic Eng.*, 125(5), 463–473.

Migniot, C. (1968). Etude des propriétés physiques de différents sédiments très fins et de leur comportement sous des actions hydrodynamiques. *La Houille Blanche*, 591–620.

Migniot, C. (1989). Tassement et rhéologie des vases. Première et deuxième parties. *La Houille Blanche*, 11–30 and 96–111.

Migniot, C. & Hamm, L. (1990). Consolidation and rheological properties of mud deposits. *Proc. 22nd Int. Conf. Coastal Eng.*, held July 2–6, 1990 in Delft, NL. ASCE, New York, 2975–2983.

Möller, J., Ohle, N., Daemrich, K-F., Zimmermann, C., Schüttrumpf, H. & Oumeraci, H. (2001). Einfluss der Wellenangriffsrichtung auf Wellenauflauf und Wellenüberlauf, 3. *FZK-Kolloquium, Planung und Auslegung von Anlagen im Küstenraum*, Hannover, 29. März 2001.

Morris, E.P., Peralta, G., Burn, F.G., Van Duren, L., Bouma, T.J. & Perez-Llorens, J.L. (2008). Interaction between hydrodynamics and seagrass canopy structure: Spatially explicit effects on ammonium uptake rates. *Limnology and Oceanography*, 53(4), 1531–1539.

Musumeci, R.E., Cavallaro, L., Foti, E., Scandura, P. & Blondeaux, P. (2006). Waves plus Currents Crossing at a Right Angle. Part I: Experimental investigation. *J. Geophysical Res.*, 111(C7), C07019, doi: 10.1029/2005JC002933.

Muttray, M. & Oumeraci, H. (2000). Wave transformation on the foreshore of coastal structures with different reflection properties. *Proc. 27th Int. Conf. Coastal Eng.*, Sydney, Australia ASCE, Vol. III, 2178–2191.

Naden, P.S., Rameshwaran, P. & Vienot, P. (2004). Modelling the influence of instream macrophytes on velocity and turbulence. *Fifth International Ecohydraulics Congress*, 12–15 September 2004, Madrid, Spain, Diego García de Jalón Lastra and Pilar Vizcaíno Martínez (Eds), 1118–1122.

Naden, P.S., Rameshwaran, P. Mountford, O. & Robertson, C. (2006). The influence of macrophyte growth, typical of eutrophic conditions, on river flow velocities and turbulence production. *Hydrological Processes*, 20, 3915–3938.

Naot, D. & Rodi, W. (1982). Calculation of secondary currents in channel flow. *Journal of the Hydraulics Division*, ASCE, 108, 948–969.

Naot, D., Nezu, I. & Nakagawa, H. (1996). Hydrodynamic behaviour of partly vegetated open channels. *Journal of Hydraulic Engineering*, 122, 625–633.

Nepf, H. (1999). Drag, turbulence and diffusion in flow through emergent vegetation. *Water Resources Research*, 35, 479–489.

Nepf, H. & Vivoni, E.R. (2000). Flow structure in depth-limited, vegetated flow. *Journal of Geophysical Research*, 105, 28, 547–557.

Nepf, H. & Ghisalberti, M. (2008) Flow and transport in channels with submerged vegetation, *Acta Geophysica*, 56(3), 753–777.

Neumeier, U. (2005). Quantification of vertical density variations of salt-marsh vegetation. *Estuarine, Coastal and Shelf Science*, 63, 489–496.

Neumeier, U. (2007). Velocity and turbulence variations at the edge of saltmarshes. *Continental Shelf Research*, 27, 1046–1059.

Newcombe, C.P. & MacDonald, D.D. (1991). Effects of suspended sediments on aquatic ecosystems. *North American Journal of Fisheries Management*, 11, 72–82.

Newe, J. (2005). *Strandprofilentwicklung unter Sturmflutseegang*. Ph.D. thesis, Dissertation, Fachbereich Bauingenieurwesen, Leichtweiß-Institut für Wasserbau, Technische Universität Braunschweig, Braunschweig, Germany, 157 S.

Nezu, D. & Rodi, W. (1982). Calculation of secondary currents in channel flow, *J. Hydraulic Div.*, ASCE, 108, 948–968.

Nielsen, P. (1983). Entrainment and distribution of different sand sizes under water waves. *J. Sed Pet*, 53(2), pp. 423–428.

Nielsen, P. (1992). *Coastal Bottom Boundary Layers and Sediment Transport*. World Scientific Publishing, Singapore, Advanced Series on Ocean Engineering, Vol. 4.

Nielsen, P. (1999). Groundwater dynamics and salinity in coastal barriers. *J. Coastal Res*, 15(3), pp. 732–740.

Nikora, V., McEwan, I., McLean, S., Coleman, S., Pokrajac, D. & Walters, R. (2007a). Double averaging concept for rough-bed open-channel and overland flows: Theoretical background. *Journal of Hydraulic Engineering, 33*, 873–883.

Nikora, V., McLean, S., Coleman, S., Pokrajac, D., McEwan, I., Campbell, L., Aberle, J., Clunie, D. & Koll, K. (2007b). Double averaging concept for rough-bed open-channel and overland flows: Application. *Journal of Hydraulic Engineering, 33*, 884–895.

Nikora, V. (2009). Hydrodynamics of aquatic ecosystems: An interface between ecology, biomechanics and environmental fluid mechanics. *River Research and Applications.*

Nikuradse, J. (1933). Strömungsgesetze in rauhen Rohren. *VDI Forschungsheft 361*, Berlin. English translation as: Laws of flow in rough pipes. Natl. Advisory Comm. Aeronautics, *Tech. Mem.* **1292**, Transl. 1950.

Nishihara, G.N. & Ackerman, J.D. (2006). The effect of hydrodynamics on the mass transfer of dissolved inorganic carbon to the freshwater macrophyte *Vallisneria Americana. Limnology and Oceanography*, **51**(6), 2734–2745.

Noda, E.K. (1972). Equilibrium beach profile scale-model relationship. *J. Waterways, Harbors and Coastal Div.*, ASCE, **WW4**, 511–528.

Nowell, A.R.M. & Jumars, P.A. (1987). Flumes – theoretical and experimental considerations for simulation of benthic environments. *Oceanography and Marine Biology Annual Review,* **25**, 91–112.

Nwogu, O.U., Mansard, E.P., Miles, M.D. & Isaacson, M. (1987). Estimation of Directional Wave Spectra by the Maximum Entropy Method. *Proc. of Wave Analysis and Generation in Lab Basins, 22nd Congress IAHR*, 363–376.

Obhrai, C. & Sutherland, J. (2009). *CoMIBBS physical model report*. HR Wallingford Report TR172, Release 2.0, February 2009. Also HYDRALAB-III Report JRA1-08-06.

O'Donoghue, T. & Clubb, G.S. (2001). Sand ripples generated by regular oscillatory flow. *Coastal Engineering, 44*, pp. 101–115.

O'Donoghue, T. & Wright, S. (2004), Flow tunnel measurements of velocities and sand flux in oscillatory sheet flow for well-sorted and graded sands, *Coastal Engineering*, **51**, 1163–1184.

Ofstad, L.J. (2002). Marine bunndyr som slambeitere i oppdrettskar vedmarin yngelproduksjon. Cand. Scient. Thesis, NTNU, Dept. of Zoology.

Ohle, N., Daemrich, K-F., Zimmermann, C., Möller, J., Schüttrumpf, H. & Oumeraci, H. (2002). The Influence of Refraction and Shoaling on Wave Run-Up under Oblique Waves. *Proc. of the 28th Inter. Conf. on Coastal Engineering, Solving Coastal Conundrums*, Cardiff, Wales, 08 July 2002.

Ohle, N., Daemrich, K-F., Zimmermann, C., Möller, J., Schüttrumpf, H. & Oumeraci, H. (2003). Wellenauf-lauf an Seedeichen, 4. *FZK-Kolloquium*, Hannover, 20. März 2003.

Old, G.H., Packman, J.C. & Scholten, H. (2005). Supporting the European Water Framework Directive: The HarmoniQuA Modelling Support Tool (MoST). **In:** Zerger, A. & Argent, R.M. (Eds.), *MODSIM 2005 International Congress on Modelling and Simulation*, Melbourne, ISBN: 0-9758400-2-9, 2825–2831, December 2005.

Ouellet, Y. & Datta, J. (1986). A survey of wave absorbers. *J. Hydraul. Res.*, 24(4), 265–280.

Oumeraci. H. (1984). Scale effects in coastal hydraulic models. *Symposium on scale effects in modelling hydraulic structures*, H. Kobus (Ed.), International Association for Hydraulic Research, 7.10.1–7.10.7.

Oumeraci, H. (1994a). *Scour in front of vertical breakwaters- Review of scaling problems*. Mitteilungen Leichtweiss-Institute, Techn. Univ. Braunschweig, 345–383.

Oumeraci, H. (1994b). *Lecture notes on physical modelling, Part I Dimensional Analysis and Scaling Laws and Part II: Physical models in coastal engineering.*

Oumeraci, H. (1999). Strength and Limitations of Physical Modelling in Coastal Engineering – Synergy Effects with Numerical Modelling and Field Measurements.

Proceedings of Hydralab-workshop "Experimental Research and Synergy Effects with mathematical Models", 7–38, Hannover, Germany.

Oumeraci, H., Zimmermann, C., Schüttrumpf, H., Daemrich, K-F., Möller, J. & Ohle, N. (2001). Influence of Oblique Wave Attack on Wave Run-up and Wave Overtopping – 3D Model Tests at NRC/ Canada with long and short crested Waves. *Report LWI No 881/FI No643/V, Forschungs-zentrum Küste (FZK)*, Hannover.

Oumeraci, H. (2009). Composite Modelling. *Paper Manuscript for CoMIBBS*, Leichtweiss-Institute for Hydraulic Engineering and Water Resources, Technical University Braunschweig, Germany.

Owen, M.W. & Allsop, N.W.H. (1983). Hydraulic modelling of rubble mound breakwaters. *Proceedings Conference on Breakwater: Design and Construction*, ICE, London.

Owen, M.W. & Briggs, M.G. (1985). Limitations of modelling. *Proceedings Conference on Breakwater '85*, ICE, Thomas Telford, London.

Paola, C., Parker, G., Sea, R., Sinha, S.K., Southard, J.B. & Wilcock, P.R. (1992). Downstream fining by selective deposition in a laboratory flume. *Science*, 258(5089), 1757–1760.

Parker, G. & Klingeman, P.C. (1982). On why gravel bed streams are paved. *Water Resources Res.*, 18, 1409–1423.

Parker, G. (1991). Some random notes on grain sorting. *Grain sorting seminar*, Ascona, Switzerland October 21–26 1991.

Parker, G. & Wilcock, P.R. (1993). Sediment feed and recirculating flumes: a fundamental difference. *J. Hydraulic Eng.*, 119, 1192–1204.

Pasche, E. & Rouvé, G. (1985). Overbank Flow with Vegetatively Roughened Flood Plains. *Journal of Hydraulic Engineering*. 111, 1262–1278.

Peakall, J., Ashworth, P. & Best, J.L. (1996). Physical modelling in fluvial geomorphology: principles, applications and unresolved issues. **In:** *The Scientific Nature of Geomorphology*, Rhoads BL and Thorn CE (eds). John Wiley & Sons: Chichester; 221–253.

Pedras, M.H.J. & de Lemos, M.J.S. (2001). Macroscopic turbulence modeling for incompressible flow through undeformable porous media. *International Journal of Heat and Mass Transfer*, 44, 1081–1093.

Peeters, E.T.H.M., Brugmans, B.T.M.J., Beijer, J.A.J. & Franken, R.J.M. (2006). Effect of silt, water and periphyton quality on the survival and growth of the mayfly *Heptagenia sulphurea*. *Aquatic Ecology*, 40, 373–380.

Peng, Z., Zou, Q., Reeve, D. & Wang, B. (2009). Parameterisation and transformation of wave asymmetries over a low-crested breakwater. *Coastal Engineering*, 56, 1123–1132.

Peterson, C.H., Summerson, H.C. & Duncan, P.B. (1984). The influence of seagrass cover on population structure and individual growth rate of a suspension-feeding bivalve, *Mercenaria mercenaria*. *Journal of Marine Research*, 42(1), 123–138.

Peterson, C.H., Luettich Jr., R.A., Micheli, F. & Skilleter, G.A. (2004). Attenuation of water flow inside seagrass canopies of differing structure, *Marine Ecology Progress Series*, 268, 81–92.

Prepernau, U., Grüne, J., Wang, Z. & Oumeraci, H. (2009). Welleninduzierte Kolkung um einen Monopile. *Proceedings of 7. FZK Kolloquium*, Hannover, Germany (in German).

Qijin, F. (1988). Separation of Time Series on Incident and Reflected Waves in Model Test with Irregular Waves. *China Ocean Eng*, 2(4), 45–60.

Rabeni, C.F., Doisy, K.E. & Zweig, L.D. (2005). Stream invertebrate community functional responses to deposited sediment. *Aquatic Science*, 67, 395–402.

Rákóczi, L. (1987). Selective erosion of noncohesive bed materials. *Geografiska Annaler* 69 A. Uppsala.

Rameshwaran, P. & Naden, P.S. (2004a). Three-dimensional modelling of free-surface variation in a meandering channel. *Journal of Hydraulic Research*, 42, 603–615.

Rameshwaran, P. & Naden, P.S. (2004b). Modelling of turbulent flow in two-stage meandering channels. *Proceedings of the Institution of Civil Engineers-Water Management*, **157**, 159–173.

Rameshwaran, P., Naden, P. & Lawless, M. (2010). Flow modelling in gravel bed rivers: Rethinking the bottom boundary condition. *Earth Surface Processes and Landforms*, (Submitted).

Rameshwaran, P. & Shiono, K. (2007). Quasi two-dimensional model for straight overbank flows through emergent vegetation on floodplains. *Journal of Hydraulic Research*, **45**(3), 302–315.

Refsgaard, J.C. & Henriksen, H.J. (2004). Modelling guidelines – terminology and guiding principles. *Advances in Water Resources*, **27**, 71–82.

Refsgaard, J.C., Højberg, A.L. Henriksen, H.J. Scholten, H., Kassahun, A. Packman J.C. & Old, G.H. (2006). Quality Assurance of the modelling process, *The XXIV Nordic Hydrological Conference*, Vingsted Centret, Denmark.

Ribberink, J.S. (1989). The large oscillating water tunnel, technical specifications and performances. *Report H 840*, Delft Hydraulics, NL.

Ribberink, J.S. & Al-Salem, A.A. (1994). Sediment transport in oscillatory boundary layers in cases of rippled beds and sheet flow. *J. Geophysical Res.*, **99**(C6), 12707–12727.

Ribberink, J. (2007). *Scale relationships for sand transport processes on the upper shoreface*. Internal communication, HYDRALAB-Project SANDS, January 2007.

Rice, S.P., Lancaster, J. & Kemp, P. (2010). Experimentation at the interface of fluvial geomorphology. Stream ecology and hydraulic engineering and the development of an effective, interdisciplinary river science. *Earth Surface Processes and Landforms*.

Rosen, R. (1989). Similitude, similarity and scaling. *Landscape Ecology*, **3**(3–4), 207–216.

Roulund, A. (2000). Three-dimensional numerical modelling of flow around a bottom-mounted pile and its application to scour. *Ph.D. Thesis*, Department of Hydrodynamics and Water Recources, Technical University of Denmark, Lyngby.

Roulund, A., Sumer, B.M., Fredsøe, J. & Michelsen, J. (2005). Numerical and experimental investigation of flow and scour around a circular pile. *J. Fluid Mech.*, **534**, 351–401.

Rousseaux, G. (2003). *Etude de l'instabilité d'une interface fluide-granulaire: Application à la morphodynamique des rides de plage*. PhD Thesis, University of Paris 6.

Rudolph, D. & Bos, K.J. (2006). Scour around a monopile under combined wave-current conditions and low KC-numbers. *Proceeding of the International Conference on Scour and Erosion (ICSE)*, Amsterdam. CD-ROM.

Sambrook-Smith, G.H. & Ferguson, R.I. (1996). The gravel-sand transition: Flume study of channel response to reduced slope. *Geomorphology*, **16**(2), 147–159.

Sánchez-Arcilla, A., Gironella, X., Vergés, D., Sierra, J.P., Peña, C. & Moreno, L. (2000). Submerged breakwaters and bars. From hydrodynamics to functional design. *Proc. of 27th Conference on Coastal Engineering, Sidney*, Australia, 1821–1835.

Sánchez-Arcilla, A., Sierra, J.P., Cáceres, I., González-Marco, D., Alsina, J.M., Montoya, F. & Galofré, J. (2006). Beach dynamics in the presence of a low crested structure. The Altafulla case. *Journal of Coastal Research*, **SI 39**, 759–764.

Sand, S.E. (1979). *Three-dimensional Deterministic Structure of Ocean Waves*. Series Paper 24, Institute of Hydrodynamics and Hydraulic Engineering (ISVA), Technical University of Denmark (DTH), DK-2800, Lyngby, Denmark.

Sand, S.E. & Lundgren, H. (1979). Three-dimensional Structure of Ocean Waves. *Proc. 2nd Int Conf on the Behaviour of Offshore Structures*, BOSS '79, **1**, 117–120.

Sand-Jensen, K., Jeppesen, E., Nielsen, K., van der Bijl, L., Hjermind, A.L., Nielsen, L.W. & Iversen, T.M. (1989). Growth of macrophytes and ecosystem consequences in a lowland Danish stream. *Freshwater Biology*, **22**, 15–32.

Sand-Jensen, K. (1998). Influence of submerged macrophytes on sediment composition and near-bed flow in lowland streams. *Freshwater Biology*, **39**, 663–667.

Santel, F., Heipke, C., Könnecke, S. & Wegmann, H. (2002). Image Sequence Matching for the Determination of three-dimensional Wave Surfaces. *Proceedings of the ISPRS Commission V Symposium*, September 2–6 2002, Corfu, Greece, **Volume XXXIV Part 5**, 596–600.

Saville, T. (1958). Wave run-up on composite slopes. *Proc. 6th Int. Conf. Coastal Eng.*, 691–699.

Schäffer, H.A., Fuchs, J.U., Hyllested, P., Mathiesen, N. & Wollesen, B. (2000). An Absorbing Multi-directional Wavemaker for Coastal Applications Coastal Engineering 2000. *Conference Proceedings, American Society of Civil Engineers*, Sydney, Australia, 981–993.

Schäffer, H.A. & Steenberg, C.M. (2003). Second-order wavemaker theory for multidirectional waves. *Ocean Engineering*, **30**, 1203–1231.

Schofield, K., Pringle, C. & Meyer, J. (2004). Effects of increased bedload on algal- and detrital-based stream food webs: Experimental manipulation of sediment and macroconsumers. *Limnology and Oceanography*, **49**, 900–909.

Scholten, H., Kassahun, A., Refsgaard, J.C., Kargas, T., Gavardinas, C. & Beulens, A.J.M. (2007). A methodology to support multidisciplinary model-based water management. *Environmental Modelling & Software*, **22**, 743–759.

Science Daily (2006). It's 2025. Where Do Most People Live? Center for Climate Systems Research, July 18 2006. Available at: http://www.sciencedaily.com/releases/2006/07/060718090608.htm

Shields, A. (1936). Anwendung der Ähnlichkeits-Mechanik und der Turbulenz-forschung auf die geschiebebewegung. Preussische Versuchsanstalt für Wasserbau und Schiffbau, **Vol. 26**, Berlin.

Shimizu, Y. & Tsujimoto, T. (1993). Comparison of flood-flow structure between compound channel and channel with vegetated zone. **In:** IAHR Congress (1993) Tokyo, **1**, (A-3-4): 97–104.

Shiono, K., Chan, T.L., Spooner, J., Rameshwaran, P. & Chandler, J.H. (2009a). The effect of floodplain roughness on flow structures, bedforms and sediment transport rates in meandering channels with overbank flows: Part I. *Journal of Hydraulic Research*, **47**(1), 5–19.

Shiono, K., Chan, T.L, Spooner, J., Rameshwaran, P. & Chandler, J.H (2009b). The effect of floodplain roughness on flow structures, bedforms and sediment transport rates in meandering channels with overbank flows: Part II. *Journal of Hydraulic Research*, **47**(1), 20–28.

Sierra, J.P., Gironella, X., Sánchez-Arcilla, A., Sospedra, J. & Alsina, J.M. (2007a). Hybrid modelling of scouring-deposition in front of a coastal structure. Journal of Coastal Research, SI 50, 364–368.

Sierra, J.P., Gironella, X. & Sospedra, J. (2007b). COMIBBS. Task 1, Format of descriptions – UPC case. Research Report LIM/AHC-07-4, 15 p.

Sierra, J.P., Gironella, X., Alsina, J.M., Oliveira, T.C.A., Persetto, V., Cáceres, I., González-Marco, D. & Sospedra, J. (2008). COMIBBS Project: Composite modelling of permeable structures. Research Report LIM/AHC-08-1, 58 p.

Sierra, J.P., González-Marco, D., Mestres, M., Gironella, X., Oliveira, T.C.A., Cáceres, I. & Mösso, C. (2009a). Numerical model for wave overtopping and transmisión through permeable coastal structures. Submitted to *Environmental Modelling & Software*.

Sierra, J.P., Gironella, X., Alsina, J.M., Oliveira, T.A.C., Cáceres, I., Mösso, C. & Mestres, M. (2009b). Physical and numerical modeling of beach response to permeable low-crested coastal structures. *Journal of Coastal Research*, **SI 56**, 1065–1069.

Simons, R.R., Whitehouse, R.J.S., MacIver, R.D., Pearson, J., Sayers, P.B., Zhao, Y. & Channell, A.R. (1995). Evaluation of the UK Coastal Research Facility. *Proc. Coastal Dynamics '95*, Gdansk, Poland, 161–172.

Sleath, J.F.A. (1990). Velocities and bed friction in combined flows. *Proc. 22nd Int. Conf. on Coastal Eng.*, ASCE, **Vol. 1**, 450–463.

Smart, G.M., Aberle, M., Duncan, M. & Walsh, J. (2004). Measurement and analysis of alluvial bed roughness. *Journal of Hydraulic Research* **42**, 227–237.

Soulsby, R.L. (1997). *Dynamics of Marine Sands*. Thomas Telford Ltd., London.

Stagonas, D. & Muller, G. (2007). Wave field mapping with particle image velocimetry. *Ocean Engineering*, **34(11–12)**, 1781–1785.

Stahl, W. (1981). Dimensional analysis in mathematical biology Part I. *Bull. Math. Biophysics*, **23**, 355.

Stahl, W. (1982). Dimensional analysis in mathematical biology Part II. *Bull. Math. Biophysics*, **24**, 81.

Stamhuis, E.J., Videler, J.J., van Duren, L.A. & Müller, U.K. (2008). Applying digital particle image velocimetry to animal generated flows: Traps, hurdles and cures in mapping steady and unsteady flows in Re regimes between 10^{-1} and 10^{5}. *Experiments in Fluids*, **33**, 801–813.

Stansberg, C.T., Krokstad, J.R. & Nielsen, F.G. (1997). Model Testing of the Slow-Drift Motion of a Moored Semisubmersible in Multidirectional Waves. In: Mansard, E. (Ed.) *IAHR Seminar – Multi-directional Waves and their Interaction with Structures*. XXVII IAHR Congress, San Francisco.

Staub, C., Svendsen, I.A. & Jonsson, I.G. (1983). Measurements of the instantaneous sediment suspension in oscillatory flow. *Progress Report No. 58*, Institute of Hydrodynamics and Hydraulic Engineering (ISVA), Technical University of Denmark, 41–49.

Stephan, U. & Gutknecht, D. (2002). Hydraulic resistance of submerged flexible vegetation. *Journal of Hydrology*, **269(1–2)**, 27–43.

Stewart, H.L. (2006). Morphological variation and phenotypic plasticity of buoyancy in the macroalga *Turbinaria ornata* across a barrier reef. *Marine Biology*, **149(4)**, 721–730.

Stokes, G.G. (1847). On the theory of oscillatory waves. *Trans. Camb. Philos. Soc.*, **8**, 441–455.

Stone, B. & Shen, H.T. (2002). Hydraulic Resistance of Flow in Channels with Cylindrical Roughness. *Journal of Hydraulic Engineering*, **128**, 500–506.

Stone, M.L., Whiles, M.R., Webber, J.A., Williard, K.W.J. & Reeve, J.D. (2005). Macroinvertebrate communities in agriculturally impacted southern Illinois streams: patterns with riparian vegetation, water quality, and in-stream habitat quality. *Journal of Environmental Quality*, **34**, 907–917.

Sumer, B.M., Fredsøe, J. & Christiansen, N. (1992). Scour around vertical pile in waves. *Journal of Waterway, Port, Coastal and Ocean Engineering*, ASCE, **118(1)**, 15–31.

Sumer, B.M. & Fredsøe, J. (1999). Hydrodynamics around cylindrical structures. Technical University of Denmark, *Advanced series on ocean engineering*, 12, World Scientific, Singapore.

Sumer, B.M., Whitehouse, R.J.S. & Tørum, A. (2001). Scour around coastal structures: a summary of recent research. *Journal of Coastal Engineering*, **44**, 153–190.

Sumer, B.M. & Fredsøe, J. (2002). *The Mechanics of Scour in the Marine Environment*. World Scientific, 522 p.

Sumer, B.M., Chua, L.H.C., Cheng, N.-S. & Fredsøe, J. (2003). The influence of turbulence on bedload sediment transport. *Journal of Hydraulic Engineering ASCE*, **129**, 585–596.

Sumer, B.M. (2007). Mathematical modelling of scour: A review. *J. Hydraulic Research*, **45(6)**, 723–735.

Sumer, B.M. & Fredsøe, J. (2001). Scour around pile in combined waves and current. *Journal of Hydraulic Engineering*, **127(5)**, 403–411.

Sutherland, J., Whitehouse, R.J.S. & Chapman, B. (1999a). Scour and deposition around detached rubble mound breakwaters. *Proc. Coastal Structures '99*, Santander, ASCE, 897–904.

Sutherland, J., Channell, A.R. & Whitehouse, R.J.S. (1999b). Design and evaluation of a sediment recirculation system for a coastal wave basin, UK Coastal Research Facility. *Technical Report TR 97*, HR Wallingford.

Sutherland, J., Peet, A.H. & Soulsby, R.L. (2004a). Evaluating the performance of morphological models. *Coastal Engineering, 51*, 917–939.

Sutherland, J., Walstra, D.J.R., Chesher, T.J. & van Rijn L.C. (2004b). Evaluation of coastal area models at an estuary mouth. *Coastal Engineering, 51(2)*, 119–142.

Sutherland, J. & Obhrai, C. (2009). *CoMIBBS composite modelling report*. HR Wallingford Report TR173, which is also HYDRALAB-III Report JRA1-08-05.

Tal, M. & Paola, C. (2007). Dynamic single-thread channels maintained by the interaction of flow and vegetation. *Geology, 35*, 347–350.

TAW, (2002). *Technical report on wave run-up and overtopping at dikes*. Technical Advisory Committee for Water Retaining Structures, Delft.

Teisson, C. & Benoit, M. (1994). Laboratory Measurement of Oblique Irregular Wave Reflection on Rubble Mound Breakwaters. *Proc 24th International Conference on Coastal Engineering* ASCE, paper 257.

Tennekes, H. & Lumley, J.L. (1972). *A first course in turbulence*, MIT Press: Cambridge, Massachusetts.

Thomson, J.R., Clark, B.D., Fingerut, J.T. & Hart, D.D. (2004). Local modification of benthic flow environments by suspension-feeding stream insects. *Oecologia, 140(3)*, 533–542.

Thorne, P.D. & Hanes, D.M. (2002). A review of acoustic measurements of small-scale sediment processes. *Continental Shelf Res., 22*, 603–632.

Thorne, P.D., Davies, A.G. & Bell, P.S. (2009). Observations and analysis of sediment diffusivity profiles over sandy rippled beds under waves, *J. Geophys. Res., 114*, C02023, doi:10.1029/2008 JC004944.

Thornton, E.B. & Guza, R.T. (1983). Transformation of wave height distribution. *J. Geophys. Res., 88(C10)*, 5925–5938.

Toorman, E., Lacor, C., Awadl, E., Heredial, M. & Widera P. (2007). Scale problems in 3d sediment-transport modelling, and suggestions to overcome them. *Colloque SHF Pollution Marine*, Paris, January.

Tørum, A., Mathiesen, B. & Escutia, R. (1979). Reliability of Breakwater Model Tests. *Proceedings of Coastal Structures, '79*, American Society of Civil Engineers, 454–469.

Triantafyllou, M.S., Hover, F.S., Techet, A.N. & Yue, D.K.P. (2005). Review of hydrodynamic scaling laws in aquatic locomotion and fishlike swimming. *Applied Mechanics Reviews, 58(4)*, 226–237.

Truelsen, C., Sumer, B.M. & Fredsøe, J. (2005). Scour around spherical bodies and self-burial. *Journal of Waterway, Port, Coastal and Ocean Engineering*, ASCE, *131(1)*, 1–13.

Twu, S.W. & Lin, D.T. (1991). On a highly effective wave absorber. *Coastal Engineering, 15*, 289–405.

Twu, S.W. & Liu, C.C. (1992). An improved arrangement for the progressive wave absorber. *Proc. 23rd Int. Conf. on Coastal Engineering*, Orlando, ASCE, 726–736.

Umeda, S., Cheng, L., Yuhi, M. & Ishida, H. (2006). Three-dimensional numerical model of flow and scour around a vertical cylinder. *Proceedings of the 30th International Conference on Coastal Engineering*, 2354–2366, San Diego, USA.

Umeda, S., Yuhi, M. & Ishida, H. (2008). Three-dimensional numerical model for wave-induced scour around a vertical cylinder. *Proceedings of the International Conference on Coastal Engineering*, 2717–2729, Hamburg, Germany.

Usseglio-Polatera, J.M., Gaillard, P. & Hamm, L. (1988). Numerical modelling of the interactive influence of wind, waves and tides on currents in shallow water. **In:** *Computer modelling in ocean engineering*, Schrefler & Zienckiewitz (Eds), Balkema, Rotterdam.

Valembois, J. (1951). Methods used at the National Hydraulic Laboratory of Chatou (France) for measuring and recording gravity waves in models, Gravity Waves. *Proceedings of the NBS Semicentennial Symposium on Gravity Waves held at the National Bureau of Standards*, June 18–20, 1951, US Government Printing Office, Washington DC.

Van Dongeren, A., Klopman, G., Reniers, A. & Petit, H. (2001). High-quality laboratory wave generation for flumes and basins. *ASCE, WAVES 2001 Conference*, San Francisco.

Van Leeuwen, P.J. & Klopman, G. (1996). A new method for the generation of second-order random waves, *Ocean Engineeering*, 23(2), 167-192.

Van Gent, M.R.A. (1995). Wave interaction with permeable coastal structures. *Communications on hydraulic and geotechnical engineering*, Faculty of Civil Engineering, TU Delft, The Netherlands, ISSN 0169-6548.

Van Gent, M.R.A. & Giarrusso, C.C. (2005). Influence of foreshore mobility on wave boundary conditions, *Proceedings WAVES*, 2005.

Van Os, A.G., Soulsby, R.S. & Kirkegaard, J. (2004). The future role of experimental methods in European hydraulic research: towards a balanced methodology. *Journal of Hydraulic Research*, 42(4), 341–356.

Van Rijn, L.C. (1993). *Principles of Sediment Transport in Rives, Estuaries and Coastal Seas.* Aqua Publications, Amsterdam.

Van Rijn, L.C. (1998). *Principles of Coastal Morphology.* Aqua Publications, Amsterdam.

Van Rijn, L.C., Walstra, D.J., Grasmeijer B., Sutherland J., Pan, S. & Sierra J.P. (2003). The predictability of cross-shore bed evolution of sandy beaches at the time scale of storms and seasons using process-based Profile models. *Coastal Engineering*, 47, 295–327.

Van Rijn, L.C. (2007). *Manual Sediment Transport Measurements in Rivers, Estuaries and Coastal Seas.* Aqua Publications, Amsterdam.

Van Rijn, L.C., Tonnon, P.K., Sanchez-Arcilla, A., Cacares, I. & Grüne, J. (2010). Scaling laws for beach and dune erosion processes. Submitted to *Coastal Engineering*.

Van Waveren, R.H., Groot, S., Schiolten, H., van Geer, F.C., Wösten, J.H.M., Koeze, R.D. & Noort, J.J. (1999). *Good Modelling Practice Handbook, STOWA Report 99–05*, Dutch Dept. of Public Works, Inst. for Inland Water Management and Waste Water Treatment report 99.036. ISBN 90-5773-056-1. 167 p.

Vellinga, P. (1986). *Beach and dune erosion during storm surges.* Ph.D. Thesis, Delft University of Technology, Delft, The Netherlands (*Publication 372*, Delft Hydraulics).

Vereecken, H., Baetens, J., Viaene, P., Mostaert, F. & Meire, P. (2006). Ecological management of aquatic plants: Effects in lowland streams. *Hydrobiologia*, 570(1), 205–210.

Viguier, J., Gallissaires, J.M. & Hamm, L. (1994a). Flume measurements of mud transport on a flat bottom under steady and alternating currents. *Report no. 5 2184 R1*, SOGREAH. G8-M Coastal Morphodynamics European project, MAST contract MS2-CT92-0027.

Viguier, J., Gallissaires, J.M. & Hamm, L. (1994b). Flume measurements of mud deposition in a transversal trench under tidal action. *Report No. 5 2184 R2*, SOGREAH. G8-M Coastal Morphodynamics European project, MAST contract MS2-CT92-0027.

Viguier, J., De Croutte, E. & Hamm, L. (2002). Hydrosedimentary studies to restore the maritime character of Mont Saint-Michel. *Proc. 28th Int. Conf. Coastal Eng.*, Cardiff, Jane McKee-Smith (ed.), World Scientific, 3285–3297.

Visser, P.J. (1986). Wave basin experiments on bottom friction due to current and waves. *Proc. 20th Int. Conf. Coastal Eng.*, 1, 807–821.

Vogel, S. (1994). *Life in Moving Fluid: The Physical Biology of Flow.* 2nd ed, Princeton University Press: Princeton, NJ.

Walters, R.A. & Plew, D.R. (2008). Numerical modeling of environmental flows using DAM: Some preliminary results, *Acta Geophysica*, 56, 918–934.

Wan, E.A. & van der Merwe, R. Haykin, S. (Eds.) (2002). The Unscented Kalman Filter. In: *Kalman Filtering and Neural Networks*, John Wiley & Sons: 221–280.

Wanek, J.M. & Wu, C.H. (2006). Automated trinocular stereo imaging system for three-dimensional surface wave measurements. *Ocean Engineering, 33(5–6)*, 723–747.

Warburton, J. & Davies, T.R.H. (1998). Use of hydraulic models in management of braided gravel-bed rivers. In: Klingeman, P.C., Beschta, R.L., Komar, P.D. & Bradley, J.B. (Eds). *Gravel-Bed Rivers in the Environment*, Water Resources Publications: Highlands Ranch, Colorado, 513–542.

Warren, C.E. & Davis G.E. (1971). Laboratory Stream Research: Objectives, Possibilities, and Constraints. *Annual Review of Ecology and Systematics, 2*, 111–144.

Weggel, J.R. (1972). Maximum Breaker Height. *J. Waterways, Harbors & Coastal Eng. Div., 98(WW4)*, 529–548.

Wells, S., Sutherland, J. & Millard, K. (2009). Data management tools for HYDRALAB – a review. *HYDRALAB Report NA3-09-02*, available from http://www.hydralab.eu/publications.asp (page accessed 08/01/10).

White, F.M. (1994). *Fluid mechanics. 3rd ed.* McGraw-Hill: New York.

Whitehouse, R.J.S. & Chesher, T.J. (1994). Seabed roughness in tidal flows – a review of existing measurements. *Report SR360*, HR Wallingford Ltd.

Whitehouse, R. (1998). *Scour at marine structures – a manual for practical application*. Thomas Telford Publications, London.

Wilcock, P.R. & Southard, J.B. (1989). Bed-load transport of mixed size sediment: fractional transport rates, bed forms, and the development of a coarse surface layer. *Water Resources Res., 25*, 1629–1641.

Wilcock, P.R. & McArdell, B.W. (1993). Surface-based fractional transport rates – mobilization thresholds and partial transport of a sand-gravel sediment. *Water Resources Res., 29(4)*, 1297–1312.

Wilcock, P.R. & McArdell, B.W. (1997). Partial transport of a sand/gravel sediment. *Water Resources Res., 33(1)*, 235–245.

Wilcock, P.R. (1998). Two-fraction model of initial sediment motion in gravel-bed rivers. *Science, 280(5362)*, 410–412.

Wilcock, P.R. (2001). The flow, the bed and the transport: interaction in flume and field. In: Mosley, M.P. (Ed), *Gravel Bed Rivers*, New Zealand Hydrological Society, Auckland, NZ, 183–209.

Wilson, C.A.M.E. & Horritt, M.S. (2002). Measuring the flow resistance of submerged grass. *Hydrological Processes, 16(13)*, 2589–2598.

Wilson, C.A.M.E., Stoesser, .T & Bates, P.D. (2005). Open channel flow through vegetation, In: Bates, P.D., Lane, S.N. & R.I. Ferguson, R.I. (Eds.) *Computational Fluid Dynamics Applications in Environmental Hydraulics*. John Wiley and Sons. 395–428.

Wolters, G., Müller, G., Bullock, G., Obhrai, C., Peregrine, H. & Bredmose, H. (2004). Field and large scale model tests of wave impact pressure propagation into cracks. *Proc. 29th Int. Conf. Coastal Engineering, Lisbon*, 4027–4039.

Wolters, G., Müller, G., Bruce, T. & Obhrai C. (2005). Large-scale experiments on wave downfall pressures, Maritime Engineering. *Proc. of the Inst. of Civil Engrs, 158(MA4)*, 137–145.

Yalin, M.S. (1989). Fundamentals of Hydraulic Physical Modelling, In:, Martins, R. (Ed.). (1989) *Recent Advances in Hydraulic Physical Modelling*. Kluwer Academic Publishers, Dordrecht, The Netherlands, pp. 567–588.

Yokoki, H. Isobe, M. & Watanabe, A. (1994). On a Method for Estimating Reflection Coefficient in Short-crested Random Seas. *Proc 24th International Conference on Coastal Engineering*, ASCE, paper 168, 384–385.

Yund, P.O. & Meidel, S.K. (2003). Sea urchin spawning in benthic boundary layers: Are eggs fertilized before advecting away from females? *Limnology and Oceanography, 48(2)*, 795–801.

Zelt, J.A. & Skelbreia, J.E. (1992). Estimating incident and reflected wave fields using an arbitrary number of wave gauges. *Proc 23rd International Conference on Coastal Engineering*, Venice, Italy, 777–788.

Zhu, L. (2008). Scaling laws for drag of a compliant body in an incompressible viscous flow. *J. Fluid Mech.*, **607**, 387–400.

Bibliography

The chapters form an introduction into the various fields of hydraulic modelling and experimentation. The following references and guidelines provide a general account:

WAVES

Aage, C. (1999). Model Testing – Bringing the Ocean into the Laboratory. *Proceedings of the 2nd HYDRALAB Workshop*, Rungsted Kyst, Denmark.

Aziz Tayfun M. (1993). Sampling-rate errors in statistics of wave heights and periods. *Journal of Waterway, Port, Coastal and Ocean Engineering*, 119(2).

Benoit, M. (1994). Extensive Comparison of Directional Wave Analysis Methods from Gauge Array Data. *Proc. 2nd Int. Symp. on Ocean Wave Measurement and Analysis*, ASCE.

Biésel, F (1951). 'Les appareils générateurs de houle en laboratoire', *La Houille Blanche*, 6 (**Nos 2, 4 et 5**).

British Standards Institution (1991). *BS 6349: Part 7. Maritime structures – Guide to the design and construction of breakwaters.*

Canning, P., She, K. & Morfett, J. (1998). Application of video imaging to water wave analysis. Proc. 3rd Int. Conf on Hydro-Science and Engineering, ICHE, Cottbus/Berlin, Germany.

Cox, C., McCleave, B., Welch, C.R., Seabergh, W. & Curtis, W. (2001). Phase profilometry measurement of wave field histories. Proc. Waves 2001, San Francisco, CA, U.S.A, 23–32.

Curtis, W.R., Hathaway, K.K., Seabergh, W.C. & Holland, K.T. (2001). Measurement of physical model wave diffraction patterns using video. *Proc. Waves 2001*, San Francisco, CA, U.S.A, 23–32.

Daemrich, K-F. & Kohlhase, S. (1978a). Wave Diffraction at Harbour Entrances with Overlapping or Displaced Breakwaters. *Proc. Intern. Conf. on Water Resources Development*, Bangkok.

Daemrich, K-F. & Kohlhase, S. (1978b). Influence of Breakwater-Reflection on Diffraction. *Intern. Conf. on Coastal Engineering*, Hamburg.

Daemrich, K-F. & Kohlhase, S. (1980). Diffraction and Reflection at Rubble-Mound Breakwaters. *Proc. Intern. Conf. on Water Resources Development*, Taipei.

Daemrich, K-F., Kohlhase, S. & Partenscky, H-W. (1983). Investigations of MACH-Reflection Including Breaking and Irregular Waves. *Proc. Intern. Conf. on Coastal and Port Engineering in Developing Countries*, Colombo.

Daemrich, K-F. (1996). Diffraktion und Reflexion von Richtungsspektren mit linearen Überlagerungs-modellen (unpublished). *Report of Franzius-Institute for Hydraulic, Waterways and Coastal Engineering*, Hannover, Germany.

Daemrich, K-F. & Mathias, H-J. (1999a). Overtopping at Vertical Walls with Oblique Wave Approach. *Fifth Intern. Conference on Coastal & Port Engineering in Developing Countries (COPEDEC V)*, Cape Town, South Africa.

Daemrich, K-F. & Mathias, H-J. (1999b). Overtopping at Vertical Walls with Oblique Wave Approach. *Proc. of the HYDRALAB-Workshop on Experimental Research and Synergy Effects with Mathematical Models*, Hannover, Germany, 17-19.2.1999, Forschungszentrum Küste (FZK).

DNV Recommended Procedure, (2009). *DNV-RP-C205, Environmental Conditions and Environmental Loads, Ch. 10 Hydrodynamic Model Testing.*

Elshoff, I.J.P., Janssen, T.T. & van Dongeren, A.R. (2001). Video Observations of Laboratory Waves. *Proc. Waves 2001*, San Francisco, CA, U.S.A, 13–22.

Gierlevsen, T., Hebsgaard, M. & Kirkegaard, J. (2001). Wave disturbance modelling in the Port of Sines, Portugal – with special emphasis on long-period oscillations, *Proc. Int. Conference on port and maritime R&D and technology (ICPMRDT)*, Singapore, 29–31 Oct. 2001. Vol. 1, 337–344.

Hughes, S.A. (1993). *Physical Models and Laboratory Techniques in Coastal Engineering*, World Scientific Publishing, Singapore – ISBN 981-02-1540-1.

HYDRALAB (2001). *Hydrodynamics and ice engineering, inventory of experimental facilities in Europe, ver. I.I*, September, 2001 www.hydralab.eu

IAHR (1987a). Wave Analysis and Generation in Laboratory Basins. *22nd IAHR Congress*, Lausanne, 1–4 September 1987, The National Research Council of Canada, 1987. - ISBN: 0-660-53811-3.

IAHR (1997). Multi-directional Waves and their Interaction with Structures. *27th IAHR Congress*, San Francisco, 10–15 August 1997. Mansard, E. (Ed.). – Ottawa: The National Research Council of Canada, 1997. - ISBN: 0660170930.

IAHR (1987b). List of Sea State Parameters. *IAHR Seminar on wave generation and analysis in basins*, Lausanne 1987 / Supplement to PIANC Bulletin No. 52, 1986.

Isobe, M., Kondo, K. & Horikawa, K. (1984). Extension of MLM for Estimating Directional Wave Spectrum. *Proc. Symp. on Modeling of Directional Seas*, paper A-6, 15 pages.

ITTC (2005a). *Recommended Procedures and Guidelines: Procedure 7.5-02-07-01.1, Laboratory Modelling of Multidirectional Irregular Wave Spectra*, Rev. 00, 2005.

ITTC (2005b). *Recommended Procedures and Guidelines: 7.5-02-07-03.1*, Floating Offshore Platform Experiments, Rev. 01, 2005.

ITTC (2005c). *The Ocean Engineering Committee, Final Recommendations to the 24th ITTC*, 2005.

Kofoed-Hansen, H., Sloth, P., Sørensen, O.R. & Fuchs, J.U. (2000). Combined numerical and physical modelling of seiching in exposed new marina. *Proc ICCE 2000. 27th Coastal Eng. Conf.*, Sydney, Australia.

Madsen, P.A., Sørensen, O. & Schäffer, H.A. (1997). Surf zone dynamics simulated by a Boussinesq type model. Part 2: Surf beat and swash oscillations for wave groups and irregular waves. *Coastal Engineering*, 32(4), 289–319.

Moxnes, S. & Larsen, K. (1998). Ultra Small Scale Model Testing of a FPSO Ship. *Proceedings of the 17th Int. Offshore Mechanics and Arctic Engineering Conference, OMAE'98*, Lisbon, OMAE 98-0381.

Schäffer, H.A. & Steenberg, C.M. (2003). Second-order wavemaker theory for multi-directional waves. *Ocean Engineering*, 30, 1203–1231.

Stansberg, C.T., Øritsland, O. & Kleiven, G. (2000). VERIDEEP: Reliable Methods for Laboratory Verification of Mooring and Stationkeeping in Deep Water. *Proceedings of the 2000 Offshore Technology Conference*, Houston, Texas, OTC 12087.

Takeda, Y. (1999). Ultrasonic Doppler method for velocity profile measurement in fluid dynamics and fluid engineering. *Journal Experiments in Fluids*, 26, 177–178.

Van Dongeren, A.R., Petit, H.A.H., Klopman, G. & Reniers, A.J.H.M. (2001). High-Quality Laboratory Wave Generation for Flumes and Basins. *Proc. Waves 2001*, San Francisco, CA, U.S.A, 1190–1199.

Zimmermann, C. & Schulz, N. (1994). Morphological effects from waves and tides on artificially stabilized forelands in the Wadden Sea. *Proceedings International Symposium on Habitat Hydraulics*, Trondheim.

Zimmermann, C. & v Lieberman, N. (1996). Morphological Effects from Waves and Tides on Artificially Stabilized Forelands in the Wadden Sea (Konferenzbeitrag). *Proceedings of the 1st International Symposium on Habitat Hydraulics*, The Norwegian Institute of Technology, Trondheim, Norway, 625–637.

BREAKWATERS

British Standards Institution (1991). British Standard Code of practice for Maritime structures, Part 7 (BS 6349). Guide to the design and construction of breakwaters. British Standards Institution, London.

CIRIA, CUR, CETMEF. (2007). The Rock Manual. The use of rock in hydraulic engineering. (2nd ed.), C683, CIRIA, London.

Dalrymple R.A. (1985). Physical modelling in coastal engineering. A.A. Balkema, Rotterdam.

Hughes S.A. (1993). Physical models and laboratory techniques in coastal engineering. Advanced Series on Ocean Engineering, Vol. 7, World Scientific Publishing, Singapore.

Owen M.W. & Allsop N.W.H. (1983). Hydraulic modelling of rubble mound breakwaters, Proceedings Conference on Breakwater: Design and Construction, ICE, London.

Owen M.W. & Briggs M.G. (1985). Limitations of modelling, Proceedings Conference on Breakwater '85, ICE, Thomas Telford, London.

TAW (2002). Technical report on wave run-up and overtopping at dikes. Technical Advisory Committee for Water Retaining Structures.

SEDIMENT DYNAMICS

Dalrymple, R.A. (1985). *Physical Modelling in Coastal Engineering*. Balkema, Rotterdam.

Fredsøe, J. & Deigaard, R. (1992). *Mechanics of Coastal Sediment Transport*. World Scientific Publishing, Singapore, Advanced Series on Ocean Engineering, Vol. 3.

Hughes, S.A. (1993). Physical models and laboratory techniques in coastal engineering. World Scientific.

HYDRALAB (2004). Strategy paper: The future role of experimental methods in European hydraulic research – towardsa balanced methodology. *J. Hydr. Research*, 42(4), 341–356.

Nielsen, P. (1992). *Coastal Bottom Boundary Layers and Sediment Transport*. World Scientific Publishing, Singapore, Advanced Series on Ocean Engineering, Vol. 4.

Van Rijn, L.C. (1993). *Principles of Sediment Transport in Rivers, Estuaries and Coastal Seas*. Aqua Publications, Amsterdam.

Van Rijn, L.C. (1998). *Principles of Coastal Morphology*. Aqua Publications, Amsterdam.

Van Rijn, L.C. (2007). *Manual Sediment Transport Measurements in Rivers, Estuaries and Coastal Seas*, Aqua Publications, Amsterdam.

Yalin, M.S. (1971). *Theory of hydraulic models*, Macmillan.

COMPOSITE MODELLING

CoMIBBS project reports (available from www.hydralab.eu):

Gerritsen, H., Caires, S. & Van den Boogaard, H.F.P. (2007). Interim report on analysis of functionality of models. *HYDRALAB-III Deliverable JRA1.1* for Consortium Participants and the EC.

Sutherland, J. (2009). Evaluation of Composite Modelling Techniques. *HYDRALAB-III Deliverable JRA1.2* for Consortium Participants and the EC.

LNEC (2008). CoMIBBS Task 2: Composite Modelling Report. *LNEC Report*. December. Lisbon, Portugal.

Van den Boogaard, H., Gerritsen, H. & Caires, S. (2008). CoMIBBS – Composite Modelling of the Interactions Between Beaches and Structures. Error correction procedure for a numerical wave model and inverse application. Deltares Report X0348, November 2008, 81 pages.

Sutherland, J. & Obhrai, C. (2009). CoMIBBS Composite Modelling Report. *HR Wallingford Report TR173, Release 2.0* (February 2009) (HYDRALAB document JRA1-08-05).

Sierra, J.P. Gironella, X. Alsina, J.M., Oliveira, T.A.C. Persetto, V. Cáceres, I. González-Marco, D. & Sospedra, J. (2008). Task3: Composite Modelling of permeable structures. *Research Report LIM/AHC-08-1*, 58 pages.

DHI (2008). CoMIBBS – *Composite Modelling of the Interactions Between Beaches and Structures, Task 4 – Composite modelling of groynes and jetties*, December 2008.

Dixen, M., Sumer, M. & Fredsøe, J. (2008). *Composite Modelling of Piles and Spheres. DTU part: Spheres*. Technical University of Denmark Department of Mechanical Engineering, Coastal, Maritime and Structural Engineering Section, July 2008.

Prepernau, U., Grüne, J. & Oumeraci, H. (2009). *Evaluation report on composite modelling of piles and spheres; Task 5 – Scour around vertical slender monopole*, Hannover, Germany.

Sumer, B.M. Dixen, M. & Sutherland, J. (2009). Strengths and weaknesses of physical and numerical models. *HYDRALAB-III Deliverable JRA1.3* for Consortium Participants and EC.

GENERAL REFERENCES

Barthel, V. & Funke, E.R. (1989). Hybrid modelling as applied to hydrodynamic research and testing. In: *Recent Advances in Hydraulic Physical Modelling*, Martins, R. (Ed.), NATO ASI Series E: Applied Sciences – Vol. 165, Kluwer Academic Publishers.

Caires, S. (2008). Investigations on the influence of a low-tide terrace using the Boussinesq-type wave model TRITON. *Deltares Report H5121*, September 2008.

Caires, S. & van Gent, M.R.A. (2009). Investigation of the influence of a low-tide terrace on wave loads using the Boussinesq-type wave mode TRITON. Abstract accepted for *Coastal Dynamics 2009*, Tokyo, September 2009 (paper in preparation).

Dee, D., Cunge, J., Labadie, G., Mateo, A.R., Mathiesen, M. Price, R., Santos, M. & Warren, R. (1994). *Guidelines for documenting the validity of computational modelling software*. IAHR bulletin, 24 p.

Frostick, L. (2008). Report on the workshop on Composite Modelling, Hull, 28 October 2008. *HYDRALAB Deliverable NA4–7*, 18 p. (available from www.hydralab.eu).

Grunnet, N., Lohier, S. & Deigaard, R. (2008a). Study of sediment bypass at coastal structures by composite modelling. *Proc. 31st Int. Conference on Coastal Engineering*, Hamburg, Germany. World Scientific, **Volume 2**, 1876–1887.

Grunnet, N., Lohier, S., Deigaard, R., Brøker, I. & Huiban, M. (2008b). Modelling of Bypass of Sediment at Harbours. *Littoral 2008: 9th international conference*, Venice.

Harmoniqua – www.harmoniqua.org

Hughes, S.A. (1993). Physical models and laboratory techniques in coastal engineering. World Scientific.

Kamphuis, J.W. (2000). Designing with models. *Proceedings 27th International Conference on Coastal Engineering*. Edge, B.L. (Ed) ASCE, 19–32.

Old, G.H., Packman, J.C. & Scholten, H. (2005). Supporting the European Water Framework Directive: The HarmoniQuA Modelling Support Tool (MoST). **In:** Zerger, A. & Argent, R.M. (Eds.), *MODSIM 2005 International Congress on Modelling and Simulation*, Melbourne, ISBN: 0-9758400-2-9, 2825-2831, December 2005.

Oumeraci, H. (1999). Strengths and Limitations of Physical Modelling in Coastal Engineering – Synergy Effects with Numerical Modelling and Field Measurements. *Proc. of the Hydralab –Workshop on Experimental Research and Synergy Effects with Mathematical Models*, Hannover, Germany, 17–19 February 1999. (Eds).: K.-U. Evers, J. Grüne, A. van Os, ISBN 3-00-004942-8, 7–38.

Oumeraci, H. (2009). Composite Modelling. *CoMIBBS report JRA1-09-10*, April 2009, 9 pages. (available from http://www.hydralab.eu/TA_publications.asp).

Refsgaard, J.C. & Henriksen, H.J. (2004). Modelling guidelines – terminology and guiding principles. *Advances in Water Resources*, **27**, 71–82.

Refsgaard, J.C., Højberg, A.L., Henriksen, H.J., Scholten, H., Kassahun, A., Packman, J.C. & Old, G.H. (2006). Quality Assurance of the modelling process, *The XXIV Nordic Hydrological Conference*, Vingsted Centret, Denmark.

Scholten, H., Kassahun, A., Refsgaard, J.C., Kargas, T., Gavardinas C. & Beulens, A.J.M. (2007). A methodology to support multidisciplinary model-based water management. *Environmental Modelling & Software*, **22**, 743–759 (available online).

Van den Boogaard, H. Gerritsen, H. & Caires, S. (2008). CoMIBBS – Composite Modelling of the Interactions Between Beaches and Structures. Error correction procedure for a numerical wave model and inverse application. *Deltares Report X0348*, November 2008, 81 pages.

Van den Boogaard, H. Gerritsen, H. Caires, S. & Van Gent, M. (2009). Composite modelling by applying an inverse technique in analysing interactions between beaches and structures, paper 188a178. *Proc. 8th Int. Conf on Hydroinformatics, Concepción*, Chile, 12–16 January, 2009, 10 pages.

Van den Boogaard, H., Gerritsen, H. Caires, S. & Van Gent, M. (2009). Wave attack on sea defences – potential benefits of a composite modelling approach, paper 10443. *Proc. 33rd IAHR Congress, Vancouver*, 10–14 August 2009, 8 p.

Van Os, A.G., Soulsby, R.S. & Kirkegaard, J. (2004). The future role of experimental methods in European hydraulic research: towards a balanced methodology. *Journal of Hydraulic Research*, **42(4)**, 341–356.

Van Steeg, P. (2007). Physical model investigations on the influence of a low-tide terrace. *Delft Hydraulics Report H4771*, December 2007, 10 p.

van Waveren, R.H., Groot, S., Schiolten, H., van Geer, F.C., Wösten, J.H.M., Koeze, R.D. & Noort, J.J. (1999). *Good Modelling Practice Handbook*, STOWA Report 99–05, Dutch Dept. of Public Works, Inst. for Inland Water Management and Waste Water Treatment report 99.036. ISBN 90-5773-056-1. 167 p.

Zhang, H., Schäffer. H.A. & Jakobsen, K.P. (2007). Deterministic combination of numerical and physical coastal wave models, *Coastal Eng*, **54**, 171–186.

For Product Safety Concerns and Information please contact our EU
representative GPSR@taylorandfrancis.com Taylor & Francis Verlag GmbH,
Kaufingerstraße 24, 80331 München, Germany

Printed and bound by CPI Group (UK) Ltd, Croydon, CR0 4YY
01/05/2025
01858473-0001